International Series in Operations Research & Management Science

Founding Editor

Frederick S. Hillier, Stanford University, Stanford, CA, USA

Volume 337

The book series **International Series in Operations Research and Management Science** encompasses the various areas of operations research and management science. Both theoretical and applied books are included. It describes current advances anywhere in the world that are at the cutting edge of the field. The series is aimed especially at researchers, advanced graduate students, and sophisticated practitioners.

The series features three types of books:

- Advanced expository books that extend and unify our understanding of particular areas.
- Research monographs that make substantial contributions to knowledge.
- Handbooks that define the new state of the art in particular areas. Each handbook will be edited by a leading authority in the area who will organize a team of experts on various aspects of the topic to write individual chapters. A handbook may emphasize expository surveys or completely new advances (either research or applications) or a combination of both.

The series emphasizes the following four areas:

Mathematical Programming: Including linear programming, integer programming, nonlinear programming, interior point methods, game theory, network optimization models, combinatorics, equilibrium programming, complementarity theory, multiobjective optimization, dynamic programming, stochastic programming, complexity theory, etc.

Applied Probability: Including queuing theory, simulation, renewal theory, Brownian motion and diffusion processes, decision analysis, Markov decision processes, reliability theory, forecasting, other stochastic processes motivated by applications, etc.

Production and Operations Management: Including inventory theory, production scheduling, capacity planning, facility location, supply chain management, distribution systems, materials requirements planning, just-in-time systems, flexible manufacturing systems, design of production lines, logistical planning, strategic issues, etc.

Applications of Operations Research and Management Science: Including telecommunications, health care, capital budgeting and finance, economics, marketing, public policy, military operations research, humanitarian relief and disaster mitigation, service operations, transportation systems, etc.

This book series is indexed in Scopus.

Daniel P. McGibney

Applied Linear Regression for Business Analytics with R

A Practical Guide to Data Science with Case Studies

 Springer

Daniel P. McGibney
Department of Management Science
University of Miami
Coral Gables, FL, USA

ISSN 0884-8289   ISSN 2214-7934   (electronic)
International Series in Operations Research & Management Science
ISBN 978-3-031-21482-0   ISBN 978-3-031-21480-6   (eBook)
https://doi.org/10.1007/978-3-031-21480-6

This Springer imprint is published by the registered company Springer Nature Switzerland AG
The registered company address is: Gewerbestrasse 11, 6330 Cham, Switzerland

For my family, friends, and students.

Preface

In regression analysis, the concepts to be learned are mathematical and require programming and intuition. The mathematical and intuition-based content can be taught in a traditional sense, but real-world scenarios are necessary to inspire students to remain interested. This book focuses on applied regression analysis for business students with an introduction to the R programming language. It emphasizes illustrating and solving real-time, up-to-date problems. Modern and relevant case studies from business are available within the text, along with clear and concise explanations of the theory, intuition, hands-on examples, and the coding required to use regression modeling. Each chapter contains the mathematical formulation and details of regression analysis and provides a practical in-depth breakdown using the R programming language.

This book stresses the practical usage of regression analysis, making it applicable to data sets encountered in everyday business. Specifically, while traditional books contain excessive theories and mathematical formulas, they can overwhelm business students without advanced mathematical knowledge. On the contrary, some books that stress intuitive lessons often omit the coding components, which prevents students from understanding the process. As such, to equip business students with an understanding of the basic principles needed to apply regression analysis without extensive and complex mathematical knowledge, this book blends theory and application by engaging students with worked-through real-life examples accompanied by detailed coding. Learning through relevant business cases is inevitable in today's business world. To that end, numerous practical applications and exercises are available in this book, allowing students to quickly apply complex math to the data sets they collect.

In this digital era, businesses have become highly dependent on data-driven analytics to guide their decision process. A survey by the GMAC even found that companies are increasingly looking to hire talent in data science and business analytics for advanced positions. Due to the increasing demand, the number of programs and courses in business analytics and data science has drastically grown in recent years. For example, Miami Herbet Business School at the University of Miami currently offers at least seven annual sections that cover regression analysis. However, no single textbook is designated for applied regression analysis in business.

Hence, the primary audience of this book will be (1) advanced undergraduate-level students studying business analytics and (2) graduate students studying business: including but not limited to students who study Master of Science in Business Analytics, Master of Science in Accounting, Master of Science in Tax, and Master of Business Administration. Nevertheless, this book suits advanced undergraduate-level students studying engineering, math, and statistics.

Coral Gables, FL, USA   Daniel P. McGibney
September 25, 2022

Acknowledgments

We would like to acknowledge the work of our reviewers, who provided wonderful suggestions and comments regarding the content and advancement of this work. Special thanks to the following:

Bryanna DeSimone, MSBA
Cibeles Duran, MBA

Ashley De Hechavarria, MBA
Heather Sasinowska, Ph.D.

Additionally, we would like to thank the following for their support for their helpful suggestions and comments which have helped to bring this textbook to print:

Sawyer Bohl, MSBA
Meghan Homer, EDD
Doug Lehmann, Ph.D.
Yixiao Li, MSBA
Jigar Patel, Ph.D.
Vincentia Yuen, MS

Alexa DeBartolo, MSA
Danielle Houck, MFA
Hangwei Li, MSBA
Anh Ninh, Ph.D.
Tessa Sutherland, BSBA

A portion of the funding for this textbook was paid for by the Deloitte Institute for Research and Practice in Analytics (DIRPA) and the Miami Herbert Business School (MHBS) at the University of Miami. We are also indebted to the staff at Springer Nature, particularly Jialin Yan (Associate Editor) and Ramya Prakash (Project Coordinator).

Daniel P. McGibney, Ph.D.

Contents

1	**Introduction**		1
	1.1	Introduction	1
	1.2	History	1
	1.3	Linear Regression, Machine Learning, and Data Science	2
	1.4	Case Studies	2
	1.5	R Versus Python	3
	1.6	R Installation	4
		1.6.1 R (Programming Language)	4
		1.6.2 RStudio IDE	4
	1.7	Book Organization	5
2	**Basic Statistics and Functions Using R**		7
	2.1	Introduction	7
	2.2	Basic Statistics	8
	2.3	Sales Calls Application: Basic Statistics	9
	2.4	Data Input and Dataframes in R	10
		2.4.1 Variable Assignment	11
		2.4.2 Basic Operations in R	11
		2.4.3 The c Function	12
		2.4.4 The data.frame Function	13
		2.4.5 The read.csv Function	13
		2.4.6 Indexing Vectors	13
		2.4.7 Indexing Dataframes	14
	2.5	Accessing the Objects of a Dataframe in R	15
		2.5.1 The Head Function	15
		2.5.2 The str Function	16
	2.6	Basic Statistics in R	16
		2.6.1 The Summary Function	16
		2.6.2 The Sum Function	17
		2.6.3 The Mean Function	17
		2.6.4 The sd Function	17

2.7   Sales Calls Application: Basic Statistics in R 18
2.8   Plotting in R 19
       2.8.1   Scatterplots 19
       2.8.2   The Plot Function 19
       2.8.3   Histograms 20
       2.8.4   The Hist Function 20
       2.8.5   Boxplots 21
       2.8.6   The Boxplot Function 21
2.9   Sales Calls Application: Plotting Using R 21
2.10   Case Study: Top Companies 25
       2.10.1   Problem Statement 25
       2.10.2   Data Description 26
       2.10.3   Scatterplots 27
       2.10.4   Histograms 29
       2.10.5   Case Conclusion 31
Problems 31

3   Regression Fundamentals 33
3.1   Introduction 33
3.2   Covariance 33
3.3   Correlation Coefficient 34
3.4   Coefficient of Determination 36
3.5   Sales Calls Application: Variable Correlations 36
3.6   Least Squares Criterion 39
3.7   Sales Calls Application: Simple Regression 41
3.8   Interpolation and Extrapolation 42
3.9   Sales Calls Application: Prediction 43
3.10   Explained Deviation 44
3.11   Case Study: Accounting Analytics 46
       3.11.1   Problem Statement 46
       3.11.2   Correlation and Scatterplot 47
       3.11.3   Linear Regression Modeling 48
       3.11.4   Audit Scenarios 49
       3.11.5   Case Conclusion 51
Problems 51

4   Simple Linear Regression 57
4.1   Introduction 57
4.2   Simple Linear Regression Model 58
4.3   Model Assumptions 58
4.4   Model Variance 59
4.5   Application: Stock Revenues 60
4.6   Hypothesis Testing 61
       4.6.1   The qt Function 62
       4.6.2   The pt Function 63
4.7   Application: Using the pt and qt Functions 64

4.8	Hypothesis Testing: Student's t-Test	66
4.9	Employee Churn Application: Testing for Significance with t	67
4.10	Coefficient Confidence Interval	69
4.11	Employee Churn Application: Confidence Interval Hypothesis Testing	70
4.12	Hypothesis Testing: F-test	72
4.13	The qf Function	72
4.14	The pf Function	73
4.15	Employee Churn Application: Testing for Significance with F...	74
4.16	Cautions About Statistical Significance	76
4.17	Case Study: Stock Betas..	76
	4.17.1 Problem Statement	76
	4.17.2 Descriptive Statistics	77
	4.17.3 Plots and Graphs...	80
	4.17.4 Finding Beta Values	82
	4.17.5 Finding All the Betas	83
	4.17.6 Recommendations and Findings	86
	4.17.7 Case Conclusion	87
	Problems	88
5	**Multiple Regression**...	91
5.1	Introduction	91
5.2	Multiple Regression Model	92
5.3	Multiple Regression Equation.......................................	92
5.4	Website Marketing Application: Modeling...........................	92
5.5	Significance Testing: t..	95
5.6	Coefficient Interpretation..	95
5.7	Website Marketing Application: Individual Significance Tests	96
5.8	Significance Testing: F	97
5.9	Multiple R^2 and Adjusted R^2	98
5.10	Website Marketing Application: Multiple R^2 and Adjusted R^2.....	100
5.11	Correlations in Multiple Regression	100
5.12	Case Study: Real Estate	101
	5.12.1 Problem Statement	101
	5.12.2 Data Description	101
	5.12.3 Simple Linear Regression Models	105
	5.12.4 Multiple Regression Model	107
	5.12.5 Case Conclusion	111
	Problems	112
6	**Estimation Intervals and Analysis of Variance**	115
6.1	Introduction	115
6.2	Expected Value	116
6.3	Confidence Interval	116

6.4   House Prices Application: Confidence Interval 117
6.5   Prediction Interval 120
6.6   House Price Application: Prediction Interval 121
6.7   Confidence Intervals verse Prediction Intervals 122
6.8   Analysis of Variance 122
       6.8.1   Mean of Squares Due to Regression 123
       6.8.2   Mean Squared Error 124
       6.8.3   The F Statistic 124
6.9   ANOVA Table 125
6.10   House Price Application: ANOVA Table 125
6.11   Generalized F Statistic 128
6.12   Case Study: Employee Retention Modeling 129
       6.12.1   Problem Statement 129
       6.12.2   Data Description 130
       6.12.3   Multiple Regression Model 131
       6.12.4   Predictions 135
       6.12.5   Case Conclusion 137
       Problems 137

7 Predictor Variable Transformations 141
7.1   Introduction 141
7.2   Categorical Variables 142
7.3   Employee Salary Application: Dummy Variables 142
7.4   Employee Salary Application: Dummy Variables 2 145
7.5   Multilevel Categorical Variables 147
7.6   Employee Salary Application: Dummy Variables with
       Multiple Levels 147
7.7   Coding Dummy Variables 149
7.8   Employee Salary Application: Dummy Variable Coding 151
7.9   Modeling Curvilinear Relationships 153
7.10   Sales Performance Application: Quadratic Modeling 153
7.11   Mean-Centering 158
7.12   Marketing Toys Application: Mean-Centering 158
7.13   General Linear Regression Model 163
7.14   Interactions 164
7.15   Marketing Toys Application: Interactions 165
7.16   Case Study: Social Media 168
       7.16.1   Problem Statement 168
       7.16.2   Data Description 168
       7.16.3   Promoter A Model 170
       7.16.4   Promoter B Model 173
       7.16.5   Combined Model 174
       7.16.6   Case Conclusion 175
       Problems 176

8   Model Diagnostics 179
  8.1   Introduction 179
  8.2   Multiple Regression Model Revisited 179
  8.3   Model Assumptions 180
  8.4   Violations of the Model Assumptions 180
  8.5   Residual Analysis.. 181
  8.6   Sales Performance Application: Residual Analysis................. 182
  8.7   Constant Variance.. 183
  8.8   Twitter Application: Residual Variance 184
  8.9   Response Variable Transformations 186
  8.10 Logarithmic Transformations 186
  8.11 Other Response Variable Transformations 188
  8.12 Box–Cox Transformation 188
  8.13 Twitter Application: Box–Cox 191
  8.14 Assessing Normality... 193
  8.15 Assessing Independence.. 194
  8.16 Outliers and Influential Observations............................. 195
  8.17 Residuals and Leverage 196
    8.17.1   Leverage 196
    8.17.2   Standardized Residuals................................... 196
    8.17.3   Studentized Residuals 197
    8.17.4   Cook's Distance 197
  8.18 Case Study: Lead Generation 198
    8.18.1   Problem Statement 198
    8.18.2   Data Description... 199
    8.18.3   Revenue by Lead Generation Method 200
    8.18.4   Revenue by Dealership 203
    8.18.5   Sales Versus Radio Ads 206
    8.18.6   Sales Versus Robocalls 208
    8.18.7   Sales Versus Emails 210
    8.18.8   Sales Versus Cold-Calls 213
    8.18.9   Recommendations and Findings 217
    8.18.10 Case Conclusion 218
  Problems 218

9   Variable Selection 221
  9.1   Introduction 221
  9.2   Parsimonious Models... 221
  9.3   Airbnb Pricing Application 222
  9.4   Assessing Model Performance 225
    9.4.1   Multiple R-Squared and Adjusted R-Squared............. 225
    9.4.2   Akaike Information Criterion 225
    9.4.3   Bayesian Information Criterion 226
    9.4.4   Mallows's C_p 226
  9.5   Airbnb Pricing Application: Model Comparison..................... 226

9.6   Backward Elimination 227
9.7   Airbnb Pricing Application: Backward Elimination 229
9.8   Forward Selection 231
9.9   Airbnb Pricing Application: Forward Selection 231
9.10   Stepwise Regression 234
9.11   Airbnb Pricing Application: Stepwise Regression 235
9.12   Best Subsets Regression 238
9.13   Airbnb Pricing Application: Best Subsets Regression 1 238
9.14   Airbnb Pricing Application: Best Subsets Regression 2 240
9.15   Stepwise and Best Subsets Regression 244
9.16   Case Study: Cancer Treatment Cost Analysis 244
        9.16.1   Problem Statement 244
        9.16.2   Data Description 245
        9.16.3   Preliminary Analysis 246
        9.16.4   Revised Analysis 248
        9.16.5   Regression Modeling 251
        9.16.6   Recommendations and Findings 262
        9.16.7   Case Conclusion 263
Problems 263

A Installing Packages 267
   A.1   Installation of the ggplot2 Package 267
   A.2   Loading in the ggplot2 Package 268
   A.3   Additional Installation Methods 268

B The quantmod Package 269
   B.1   Data Source 269
   B.2   Downloading a Single Stock or Index 269
   B.3   Plotting Stock Prices 271
   B.4   Multiple Stock Download 271
   B.5   Calculate Stock Returns 272
   B.6   Create a Dataframe 272

Bibliography 275

About the Author

Dr. Daniel McGibney is an Assistant Professor of Professional Practice at the University of Miami Herbert Business School. He currently teaches Analytics to both graduate and undergraduate students. Over the years, he taught various classes covering Analytics and Data Science, ranging from Basic Statistics to Big Data Analytics and Deep Learning. He has taught Applied Linear Regression Analysis to students pursuing their MSBA, MBA, MST, and MAcc. He also actively oversees and mentors the graduate capstone projects in Analytics for MSBA students, collaborating with Deloitte, Visa, Carnival, Citi, Experian, and many other companies. Moreover, Dr. McGibney served as the Program Director for the MSBA degree program. He advised students, oversaw admissions, expanded industry partnerships, and advanced the program's curriculum during his tenure as program director.

Before joining the Miami Herbert Business School, Dr. McGibney was a Visiting Assistant Professor of Mathematics at the College of William and Mary. He taught and researched Statistics, Analytics, and Machine Learning. He received his Ph.D. in Systems Science and Mathematics from Washington University in St. Louis and has Master's degrees in Engineering and Mathematics from Washington University in St. Louis and Southern Illinois University Edwardsville. He also has industry experience in Engineering, Simulation, Regression Modeling, Supply Chain, and Optimization from working as an Operations Research Analyst for Northrop Grumman and an Instrument and Control Engineer at NooterEriksen.

Chapter 1
Introduction

In God we trust. All others must bring data.

—*W. Edwards Deming*

1.1 Introduction

Business analytics uses modern computing methods to report, enhance, and provide insights into modern businesses. Regression analysis does these actions through data by predicting unknown values, assessing differences among groups, and checking the relationships among variables. When regression analysis is applied to the right data set in the right way, the results can make businesses extremely profitable, whether the objective is predicting the sale price of houses, assessing marketing methods, or predicting the number of likes on a social media post. This book has countless applications of business examples where regression analysis produces valuable insights. This chapter begins with a discussion of the history of regression analysis and its role in data science, machine learning, and artificial intelligence (AI). Also, we will provide an overview of each of the eight case studies in this book. These cases offer detailed analyses of how to use regression analysis to obtain actionable business findings.

1.2 History

Regression analysis, developed over two hundred years ago, has a rich history. French mathematician Adrien-Marie Legendre published the first writings on regression in 1805, and German mathematician Johann Carl Friedrich Gauss in 1809. Their work focused on astronomical applications. Later that century, Sir Francis Galton coined the term "regression" to describe the phenomenon where the descendants of tall ancestors tend to have lower heights. Specifically, the name regression stems from the concept that the descendants' heights had a "regression

© The Author(s), under exclusive license to Springer Nature Switzerland AG 2023
D. P. McGibney, *Applied Linear Regression for Business Analytics with R*,
International Series in Operations Research & Management Science 337,
https://doi.org/10.1007/978-3-031-21480-6_1

toward the mean." Galton's student, Karl Pearson, worked with Udny Yule to put regression into a more general statistical framework. Another notable pioneer, Ronald Fisher, continued the statistical work, developed the analysis of variance (ANOVA) methodology, and set the foundations for modern statistics. While regression analysis has roots in astronomy and biology, there are many modern applications in almost every academic field.

1.3   Linear Regression, Machine Learning, and Data Science

In modern statistics, the fundamental concepts of linear regression have changed little over the past decades. More work has been done to develop and expand on these existing concepts, and regression continues to be an active area of research. Today, many linear regression methods are broadly considered machine learning or, even more broadly, artificial intelligence.

Traditional linear regression analysis can be used to make predictions and inferential understanding. Machine learning aims to make predictions, sometimes using a "black-box" approach and typically not used for inference. Despite this pitfall, machine learning is gaining popularity because predictions are needed with the copious amounts of data being collected in this modern age.

Data scientists are required to have statistical, programming, and domain knowledge. To become a data scientist or develop expertise in business analytics, the reader must learn many powerful tools to understand and manipulate data. Linear regression and the R programming language are a few of these powerful tools.

1.4   Case Studies

The case studies in this book aim to show how analytics can solve numerous problems, particularly in business domains. Each case covers a different business application and demonstrates the relevance analytics can play in each. The cases are listed here:

1. Top 200 Companies
   In this case study, we will create visualizations and generate summary statistics from a small set of the world's top 200 companies.
2. Accounting Analytics
   Using a data set consisting of the adjusted gross income and the itemized deductions for a select list of subjects, we look at the relationship between the variables and investigate any irregularities.

3. Stock Betas
   This data set consists of monthly returns of seven stocks and the S&P 500. From these data, we will calculate the stock "betas," which are valuable measures of the relationship between stocks and a market benchmark.
4. Housing Market Analysis
   In this case study, we consider a data set of 90 cities and investigate the housing market for each city. In addition, we test a claim about the impact tax rate has on house prices.
5. Employee Retention
   This case investigates two data sets: a historical data set of employee retention records and another consisting of new applicants to the company. With these data sets, we predict the retention of new applicants and evaluate two potential employee staffing companies.
6. Social Media Analysis
   This case considers video posts advertised by one of two possible promoters. Here we analyze which promoter we should use and how the sentiment and the age of a video affect the number of likes.
7. Lead Generation
   Using a data set consisting of the performance of 142 car dealerships across the country, we analyze the data to increase sales revenue. In particular, the dealerships could use radio ads, robocalls, emails, or cold calls as a means by which to increase sales. Here we compare and contrast the different methods and recommend which method to use.
8. Cancer Treatment Cost Analysis
   In this case study, we focus on the factors of health care costs that an insurance company can use to determine premiums. This case examines the data on charges billed by health insurance for treating different types of cancer.

1.5   R Versus Python

Business analytics and data science problems are solved using various software, most notably R and Python. R and Python are relatively new compared to older programming languages such as C, C++, and Fortran. They also both offer a great deal of flexibility in coding data. One key advantage of R over Python is the number of functions available in "base R." For example, to load a data file into R, one can use the `read.csv` command, but in Python, it is necessary to use the Pandas library. These functions from base R make a difference in coding with relative ease.

Python is a general-purpose programming language not necessarily meant for statistics, even though it has significant libraries devoted to statistics and data science. Being a general-purpose programming language is a perk in many ways, and the data science community has targeted Python as a popular choice for data science. The `sklearn` library within Python is rich with machine learning algorithms. For deep learning and natural language processing, Python is a popular choice. However, for linear regression analysis, we recommend R.

R was created for statistical computing, making it a natural choice to do regression, logistic regression, time series, machine learning, and general statistical analysis. In the case of linear regression, base R is sufficient to perform the analysis and is quite powerful in doing so. In this book, base R is used more frequently than functions from R packages, but a select few packages are used in a few cases. Other key advantages include R's usage of Posit (formerly RStudio), and the reporting tools available in R. R provides some great tools for the analyst to get going on analyzing data sets. While Python offers many advantages as well, for regression analysis, we recommend using R over Python. It is highly recommended to the reader that after studying this text and programming examples, they continue their studies in both R and Python, as both have advantages and limitations.

1.6   R Installation

In order to install R, it is highly recommended that you install both R and Posit. While R is the programming language that can do R calculations, Posit is a popular interface for R referred to as an Integrated Development Environment (IDE). While installing R is sufficient for learning regression analysis in R, a better experience can be had by using R with Posit.

1.6.1   R (Programming Language)

Installing R can be done by navigating to https://www.r-project.org/ and selecting the "CRAN" (Comprehensive R Archive Network) link. This link will bring you to a list of locations from which you can download R. While the location selected is not significantly important, you may want to choose the "0-Cloud" option that will automatically direct you to servers worldwide by Posit. Next, choose the operating system of your computer, and select the most up-to-date release of R. Finally, run the downloaded file specifying the download options of your choice.

1.6.2   RStudio IDE

After installing R, the RStudio integrated development environment (IDE) can be installed by navigating to https://posit.co/. While we recommend the desktop version of RStudio, other versions can also be selected. Run the downloaded file, specifying the download options of your choice. Mac users may be prompted to move the RStudio icon into the Applications folder, which should be done as specified by the program.

Once the user is familiar with programming in R, it is further recommended that the reader explores R Markdown and Latex for generating reports.

1.7 Book Organization

Chapter 2 discusses some basic concepts in statistics and basic concepts in R programming. The following two chapters, Chaps. 3 and 4, provide the detailed calculation and theory of simple linear regression or regression analysis with only one predictor variable. Then, Chap. 5 introduces multiple regression or regression with multiple predictor variables. Multiple regression is followed by Chap. 6, a chapter on additional theory and methods, particularly analysis of variance for regression, and prediction/confidence intervals. Then, Chap. 7 provides a detailed discussion on predictor variable transformations, including dummy variables and squared predictor variables. Chapter 8 follows with a discussion of diagnosing regression models. Lastly, there is a chapter that gives an overview of variable selection methods for multiple regression models (Chap. 9).

Chapter 2
Basic Statistics and Functions Using R

Numerical quantities focus on expected values, graphical summaries on unexpected values.
—John Tukey

2.1 Introduction

Data science represents a multifaceted discipline, since it requires knowledge from statistics to understand the data, knowledge from programming to manipulate the data, and the know-how to explain the data, which is often best done with one or more visualizations. Beyond statistics as the general subject matter within the branch of mathematics, the word "statistics" carries a second definition referring to the numeric values that summarize a sample, such as the mean, median, standard deviation, and variance. The R programming language signifies a preferred choice among statisticians and data scientists to easily manipulate data and provide useful statistics on that data. R has many popular plots, but here we will focus on three of the most basic ones, which are necessary for the study of linear regression.

This chapter provides a basic overview of statistics, programming, and plotting. We review the aforementioned numeric quantities and a few other statistics. We introduce data manipulation in R at a basic level so that, after the reader understands the basic statistics, he or she can input the data and calculate the values within R. In addition, plots can provide graphical summaries of data samples and also offer important tools to the field of statistics. We cover a few basic plots and how to create them using R. The concepts are put into context with a simple application using data from a call center, which observes sales and the corresponding number of calls. In the final discussion of the chapter, we present and solve a case study utilizing a data set consisting of the world's top 200 companies. This case study uses R to generate descriptive graphs and basic statistics to understand the data.

© The Author(s), under exclusive license to Springer Nature Switzerland AG 2023
D. P. McGibney, *Applied Linear Regression for Business Analytics with R*,
International Series in Operations Research & Management Science 337,
https://doi.org/10.1007/978-3-031-21480-6_2

2.2 Basic Statistics

In basic statistics, the mean is used as a measure of center to summarize a random variable. If the random variable is X with n values in a sample $(x_1, x_2, \ldots, x_n)$, then the sample mean, denoted as $\bar{x}$, can be calculated from

$$\bar{x} = \frac{\sum_{i=1}^{n} x_i}{n}, \tag{2.1}$$

where x_i is the ith entry in X.

The variance of X is the expectation of the squared deviation of X from its mean. The variance measures the spread of a random variable, or how dispersed observations are in the distribution of the random variable. The sample variance of a random variable X is denoted as s^2 and calculated:

$$s^2 = \frac{\sum_{i=1}^{n} (x_i - \bar{x})^2}{n - 1}. \tag{2.2}$$

For computational purposes, calculating the sample variance from Eq. (2.2) is typically less convenient than calculating the variance from the mathematically equivalent equation:

$$s^2 = \frac{\sum_{i=1}^{n} x_i^2 - \left(\sum_{i=1}^{n} x_i\right)^2 / n}{n - 1}. \tag{2.3}$$

The denominator of Eqs. (2.2) and (2.3) is the degrees of freedom, which denotes the number of choices or parameters that are available in determining s or s^2. Since the value of $\bar{x}$ is known and used to calculate s^2, the number of x_i values needed is $n - 1$. More specifically, in Eq. (2.1), the value on the left-side is known, and therefore, only $n - 1$ values of x_i are needed because the remaining unknown x_i values can be found using simple algebra. It may seem odd that the denominator of the variance is $n - 1$ and not n; however, it should be noted that dividing by $n - 1$ denotes a better estimate of the unknown population variance.

The numerator of Eqs. (2.2) and (2.3) is commonly called the sum of squares of X that represents the total variation in X. Pay special attention to this value for it is the foundation for many formulas in regression analysis. The formula for the sample sum of squares is

$$SS_{xx} = \sum_{i=1}^{n} (x_i - \bar{x})^2, \tag{2.4}$$

but it is often more convenient to calculate the sum of squares using the form:

$$SS_{xx} = \sum_{i=1}^{n} x_i^2 - \left(\sum_{i=1}^{n} x_i\right)^2 / n. \tag{2.5}$$

One popular measure of variation is the sample standard deviation, which can be found by simply taking the square root of the sample variance:

$$s = \sqrt{s^2} \tag{2.6}$$

$$s = \sqrt{\frac{\sum_{i=1}^{n}(x_i - \bar{x})^2}{n-1}}. \tag{2.7}$$

Since the SS_{xx} is used to find the variance, it can also be used to find the standard deviation directly:

$$s = \sqrt{\frac{SS_{xx}}{n-1}}. \tag{2.8}$$

The following application captures the utility of these simple concepts.

2.3 Sales Calls Application: Basic Statistics

Jordan manages retirement accounts for Enright Associates. She documented the number of sales calls that she made each day for the last 6 days.

Using Table 2.1, describe the distribution by doing the following with functions in R:

(a) Compute the mean.
(b) Compute the sum of squares.
(c) Compute the variance.
(d) Compute the standard deviation.

Solution

(a) Plugging the values of X into Eq. (2.1), the sample mean is calculated to be

$$\bar{x} = \frac{\sum_{i=1}^{n} x_i}{n}$$

$$\bar{x} = \frac{12 + 18 + 5 + 25 + 15 + 8}{6}$$

$$\bar{x} = 13.83.$$

Table 2.1 Jordan's calls

Variable						
Calls (X)	12	18	5	25	15	8

(b) The sum of all the values and the sum of all the x squared values will need to be calculated to find SS_{xx}:

$$\sum_{i=1}^{n} x_i = 12 + 18 + 5 + 25 + 15 + 8 = 83$$

$$\sum_{i=1}^{n} x_i^2 = 12^2 + 18^2 + 5^2 + 25^2 + 15^2 + 8^2 = 1407.$$

Knowing these two values, Eq. (2.5) can now be utilized:

$$SS_{xx} = \sum_{i=1}^{n} x_i^2 - \left(\sum_{i=1}^{n} x_i\right)^2 / n$$

$$SS_{xx} = 1407 - (83)^2 / 6 = 258.83.$$

(c) Since the numerator from Eq. (2.3) is SS_{xx}, the variance is easily solved using Eq. (2.3):

$$s^2 = \frac{\sum_{i=1}^{n} x_i^2 - \left(\sum_{i=1}^{n} x_i\right)^2 / n}{n-1}$$

$$s^2 = \frac{258.833}{5} = 51.767.$$

(d) The standard deviation can be found by taking the square root of the variance, as in Eq. (2.6):

$$s = \sqrt{s^2}$$

$$s = \sqrt{51.767} = 7.195.$$

2.4   Data Input and Dataframes in R

To use data in R, one must either input data or load data. Some simple functions that accomplish this task are the c function and the data.frame function.

2.4.1   Variable Assignment

Assigning variables within R can be done using the equal sign. For instance, assigning the variable x to a value of 24 is done with an equal sign as shown here.

```
x = 24
```

The resulting variable is a scalar named x. Alternatively, variable assignment can be done without the equal sign.

```
x <- 24
```

Using either of the two assignment methods shown above creates a scalar value named x.

Note that x can be printed by simply entering x at the command line of the console.

```
x
```

```
## [1] 24
```

Further note that the output returned is [1] followed by the value of x (24). The [1] simply refers to the index of the output that can be ignored in this case since x is a scalar.

2.4.2   Basic Operations in R

By simply typing numbers and operations in the R console, R will perform the specified mathematical operations. Addition and subtraction are carried out with the + and − characters, respectively.

```
8 + 4 - 3
```

```
## [1] 9
```

Multiplication, division, and exponentiation are carried out with *, /, and ^ characters, respectively.

```
10 ^ 2 - 1
```

```
## [1] 99
```

Parentheses can be used to specify the order of operations.

```
10 ^ (2 - 1)
```

```
## [1] 10
```

From the previously created scalar x, we can subtract a value of 12.

```
x - 12
```

```
## [1] 12
```

If we were to define a scalar variable y with a value of 12 and subtract y from x, it would yield the same result as above.

```
y = 12
x - y
```

```
## [1] 12
```

2.4.3   The c Function

The c function is a simple function that combines data. For instance, the numbers 1, 2, and 3 can be combined into a vector called v1 by doing the following.

```
v1 = c(1, 2, 3)
```

The c function, however, is not limited to numeric data. The letters a, b, and c can be combined into a character string vector v2.

```
v2 = c("a", "b", "c")
```

Numeric and character values can be brought together into a vector as well.

```
v3 = c("a", 2, "c", 99)
```

Similar to scalar values, vectors can be printed in the console by typing the vector name. To print v1, we write the following.

```
v1
```

```
## [1] 1 2 3
```

2.4.4   The data.frame Function

The data.frame function can be used to combine vectors into a single dataframe. Since vectors v1 and v2 are the same length, a dataframe df can be created consisting of both vectors.

```
df1 = data.frame(v1, v2)
```

The dataframe can be created with the names var1 and var2. To specify the variable names accordingly, the following syntax is utilized.

```
df2 = data.frame(var1 = v1, var2 = v2)
```

2.4.5   The read.csv Function

The read.csv function can be used to load in data from a csv or other text files. If a data file "test.csv" is in the working directory, then we can create a new dataframe in the following way.

```
df3 = read.csv("test.csv")
```

To ensure that R can find the csv file, we recommend that the csv file be located in the same directory as the r or rmd file. Alternatively, the working directory can be set in RStudio or using the setwd function.

2.4.6   Indexing Vectors

Indexing is accomplished in R by using square brackets "[]" immediately following the data object. Specifically, the vector v2 has 3 values, the first of which is a. This first value can be referenced by typing:

```
v2[1]
```

```
## [1] "a"
```

The second and third values can be referenced as

```
v2[2:3]
```

```
## [1] "b" "c"
```

or by using the c function.

```
v2[c(2, 3)]
```

```
## [1] "b" "c"
```

We can also omit the first value by using a minus sign in front of the index.

```
v2[-1]
```

```
## [1] "b" "c"
```

The above code omits the first value in the vector.

2.4.7 Indexing Dataframes

The square brackets can also be used to index dataframes. Dataframes have both rows and columns that are referenced within the square brackets and separated by a comma.

```
df1[row, column]
```

To specify the first row and the second column of df1, we would specify:

```
df1[1, 2]
```

```
## [1] "a"
```

To specify all of the row values in the first column of df1, we could leave the row value empty.

```
df1[, 1]
```

```
## [1] 1 2 3
```

We could also specify the second row from both columns by leaving the column value empty.

```
df1[2, ]
```

```
##   v1 v2
## 2  2  b
```

Omitting the second row can be done using the minus sign.

```
df1[-2,]
```

```
##    v1 v2
## 1  1  a
## 3  3  c
```

2.5   Accessing the Objects of a Dataframe in R

The dataframes contain objects within them. In df1, the object v1 can be accessed using the $ symbol.

```
df1$v1
```

Recall that df2 had the vectors renamed to be var1 and var2. Accessing the first vector from df2 can also be accomplished using the $ symbol.

```
df2$var1
```

The dataframe can also be appended easily. For instance, suppose we have the vector v4 that consists of the numbers 4, 5, and 6. To append the vector v4 to df1, we could input the code below.

```
v4 = c(4, 5, 6)
df1$v4 = v4
```

2.5.1   The Head Function

While a dataframe can be printed to the console simply by typing the name of the dataframe, the head function can be a convenient way to observe the data. Specifically, the head function prints out the first 6 rows of the dataframe.

```
head(df1)
```

```
##    v1 v2 v4
## 1  1  a  4
## 2  2  b  5
## 3  3  c  6
```

Observing the first 6 rows can reveal much about the dataframe without overwhelming the console with the entire data set.

2.5.2   The str Function

The str function returns the structure of an object. The structure of v1 shows that the vector is numeric with three values and the first few values are 1, 2, and 3.

```
str(v1)
```

```
## num [1:3] 1 2 3
```

Applying the str to the dataframe df1 not only shows the details of v1, but also shows the details of v2 and v4 since they are also within the dataframe.

```
str(df1)
```

```
## 'data.frame':    3 obs. of  3 variables:
## $ v1: num  1 2 3
## $ v2: chr  "a" "b" "c"
## $ v4: num  4 5 6
```

The output of the str function reveals that v2 is a vector of character (chr) strings.

2.6   Basic Statistics in R

In the previous sections, we discussed a few of the different types of data objects that can be used in R, particularly the scalar, vector, and dataframe objects. In this section, we cover how to manipulate these objects to calculate some basic statistics.

2.6.1   The Summary Function

The summary function can also help the user to understand the data of interest. The summary of a dataframe will return the minimum, 25th percentile, median, 75th percentile, the maximum, and the mean for each numeric vector in the dataframe.

```
summary(df1)
```

```
##       v1            v2                   v4
## Min.   :1.0   Length:3           Min.   :4.0
## 1st Qu.:1.5   Class :character   1st Qu.:4.5
## Median :2.0   Mode  :character   Median :5.0
```

```
##  Mean    :2.0              Mean    :5.0
##  3rd Qu.:2.5               3rd Qu.:5.5
##  Max.    :3.0              Max.    :6.0
```

The summary function will function in a similar manner when applied to a numeric vector. The summary function can also be used for summarizing regression objects as discussed in the next chapter.

2.6.2   The Sum Function

The sum function will add up all of the values in a vector. This function can be applied to vector v1,

```
sum(v1)
```

```
## [1] 6
```

or within dataframes using the dollar sign ($) as follows.

```
sum(df1$v1)
```

```
## [1] 6
```

If all of the values in df1 were numeric, the sum function could be used to sum up all of the values in the dataframe. To get the column sums of all the numeric columns in a dataframe, one could employ the colSums function.

2.6.3   The Mean Function

The mean function calculates the mean or average of the vector.

```
mean(v1)
```

```
## [1] 2
```

Similar to the colSums function, the colMeans function can return the means of every column for a dataframe that consists of numeric values.

2.6.4   The sd Function

The sd function will calculate the standard deviation of a vector.

```
sd(v1)
```

```
## [1] 1
```

2.7   Sales Calls Application: Basic Statistics in R

Using Jordan's data from the previous application (Table 2.1), describe the distribution by doing the following with functions in R:

(a) Compute the mean.
(b) Compute the sum of squares.
(c) Compute the variance.
(d) Compute the standard deviation.

Solution

(a) Before computing the mean, the data must be entered using the c function. Here we refer to the calls variable as X.

```
X = c(12, 18, 5, 25, 15, 8)
```

The mean function can be used here.

```
mean(X)
```

```
## [1] 13.83333
```

(b) The sum of squares can be done by squaring each value of X individually, but R allows for element-wise operations. Squaring each value within the vector can be accomplished using the ^ operator.

```
sum(X^2)
```

```
## [1] 1407
```

(c) Computing the variance can be done using the var function, or by squaring the result from the sd function.

```
sd(X)^2
```

```
## [1] 51.76667
```

(d) The standard deviation can be computed using the sd function.

```
sd(X)
```

```
## [1] 7.194906
```

2.8 Plotting in R

While R is capable of advanced plotting using a package such as `ggplot2`, the functions within the base distribution of R are very capable. Throughout this text, examples will be given in base R, whereas some of the advanced graphics will be done using `ggplot2`.

2.8.1 Scatterplots

Scatterplots are an important type of plot to visually demonstrate the relationship between two numeric variables. Since the goal of regression analysis is to determine the relationship between numeric variables, the scatterplot is of particular interest to analysts performing a regression. In particular, scatterplots can be used to assess the quality of a regression as discussed in Chap. 8.

Drawing a scatterplot consists of a simple process whereby the x-axis and y-axis are drawn and then points are plotted. To generate a scatterplot in R, the `plot` function can be used.

2.8.2 The Plot Function

The `plot` function is a popular method for generating plots within R. In most instances, one uses the `plot` function to generate scatterplots. It is possible to specify only a vector within the `plot` function. While a scatterplot is generated with only a vector as the input, the plot generated places the index (or row number) on the x-axis that is typically not desirable. Here we demonstrate how to plot `v1` using the plot command assuming the `v1` value is created and falls within the R environment.

```
plot(v1)
```

Here, we show how to generate the same plot if we would like to access `v1` within the `df1` dataframe.

```
plot(df1$v1)
```

The plots above produce identical results with the exception of the y-axis label. The y-axis label becomes "v1" and "df$v1" for both of the respective plots. If we would like to change the y-axis label for the last plot, we could use the `ylab` argument within the `plot` function. The code below produces results identical to the output for `plot(v1)`.

```
plot(df1$v1, ylab = "v1")
```

The following code produces a scatterplot that has v1 on the *x*-axis and v4 on the *y*-axis assuming the vectors are within the global environment.

```
plot(v1, v4)
```

To plot v1 and v4 that are contained within df1, the following code can be used.

```
plot(df1$v1, df1$v4)
```

As discussed, when plotting the vector v1 by itself, the *x*-label and *y*-label may not be the desired names. Therefore, we can modify the plot by specifying xlab and ylab. The main argument allows us to add a title to the plot.

```
plot(df1$v1, df1$v4,
     xlab = "Vector 1", ylab = "Vector 2", main = "Title")
```

2.8.3 Histograms

A histogram refers to an important plot that analysts use to identify the shape of a distribution. Most notably, a frequency histogram separates numeric values into bins of equal length, and a bar is drawn for each bin where the height of the bar is determined by the frequency of values within the corresponding bin. The common bell-shaped normal curve represents one popular shape identified by histograms. Like the scatterplot, histograms can be used in regression analysis to verify the validity of a regression model, as discussed in Chap. 8.

2.8.4 The Hist Function

To create a histogram in base R, the hist function can be used. To get a histogram of the vector v1 contained in the global environment, we code the following.

```
hist(v1)
```

To get a histogram of the vector v1 contained within df1, we code the following.

```
hist(df1$v1)
```

Similar to the plot function, the xlab and main arguments can be used here. While the ylab argument can also be used here, the *y*-axis label defaults to "Frequency" that is typically a satisfactory label.

```
hist(df1$v1, main = "Histogram Title", xlab = "Vector 1")
```

2.8.5 Boxplots

A boxplot, sometimes called a box-and-whiskers plot, is great for visualizing the variability and the center of variables simultaneously. The variability of each variable can be observed by noting the total height of each box and whiskers and noting the height of the box. The height of each box is referred to as the interquartile range (IQR) and can be calculated by finding the 75th percentile (Q_3) and subtracting the 25th percentile (Q_1):

$$IQR = Q_3 - Q_1. \tag{2.9}$$

If observations fall outside of $Q_1 - 1.5(IQR)$ and $Q_3 + 1.5(IQR)$, then they are deemed outliers, displayed in the boxplot as unfilled circles above or below the whiskers in the standard `boxplot` function. Also, the median, a measure of center, can be observed by a black line within each box.

2.8.6 The Boxplot Function

The `boxplot` function produces a boxplot. To get a boxplot of a vector, simply specify the vector within the `boxplot` function.

```
boxplot(df1$v1)
```

To get a boxplot of a dataframe with all numeric values, specify the dataframe as the first argument. This will produce a boxplot for each of the vectors within the dataframe together on the same plot. For instance, assume we have a dataframe with all numeric values called `df`. A boxplot can be generated using the following code.

```
boxplot(df)
```

2.9 Sales Calls Application: Plotting Using R

Jordan, from the previous application, also collected the number of sales that she made each day for the last 6 days, denoted as Y.

Using Table 2.2, investigate the relationship between X and Y by doing the following:

(a) Create vectors for X and Y.
(b) Create a dataframe containing calls and sales.
(c) Use the `plot` function to create a scatterplot of calls and sales. Interpret the plot.

Table 2.2 Jordan's calls and sales

Variable						
Calls (X)	12	18	5	25	15	8
Sales (Y)	10	15	4	21	14	6

(d) Use the `hist` function to create histograms of both calls and sales. Interpret the plots.
(e) Use the `boxplot` function to create a boxplot of calls and sales on separate plots. Interpret the plots.
(f) Use the `boxplot` function to create a boxplot of calls and sales on the same plot. Interpret the plot.

Solution

(a) Vectors are created using the c function.

```
X = c(12, 18, 5, 25, 15, 8)
Y = c(10, 15, 4, 21, 14, 6)
```

(b) Using the X and Y vectors as inputs of the `data.frame` function, we create a dataframe called `df`. Here we label X as `Calls` and Y as `Sales`.

```
df = data.frame(Calls = X, Sales = Y)
```

(c) Using the `plot` function, we input the X and Y vectors. Here we label the axes and give the plot a title.

```
plot(X, Y, xlab = "Calls", ylab= "Sales", main = "Scatterplot")
```

From the scatterplot in Fig. 2.1, we note a clearly defined relationship between calls and sales. In particular, as the number of calls increases, the number of sales also increases. A positive linear trendline could be used to accurately describe the relationship between the two variables.

(d) Here we use the `hist` function to create a histogram of calls and a histogram of sales.

```
hist(X, xlab= "Calls", main = "Histogram of Calls")
hist(Y, xlab= "Sales", main = "Histogram of Sales")
```

From the left-side histogram of Fig. 2.2, we see that the highest frequency bars correspond to between 5 and 10 calls and also between 10 and 15 calls. The bins here each have frequencies of 2. The remaining bins have frequencies of 1 each. The frequencies appear to be decreasing as calls increase, and most of the days have less than 15 calls.

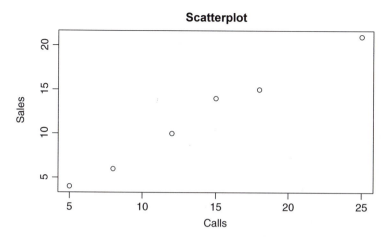

Fig. 2.1 Jordan's scatterplot

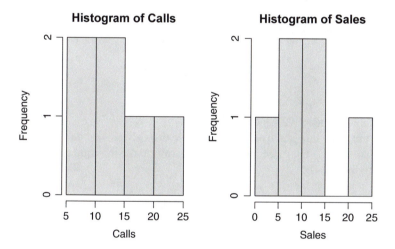

Fig. 2.2 Jordan's histograms

From the right-side histogram of Fig. 2.2, the highest frequency bars correspond to between 5 and 10 sales and also between 10 and 15 sales. Here a majority of the sales values are less than or equal to 15 with the exception of one value that is between 20 and 25.

(e) Plugging in the X and Y variables into the boxplot function individually, we have the following.

```
boxplot(X, ylab = "Calls")
boxplot(Y, ylab = "Sales")
```

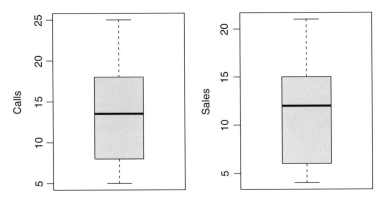

Fig. 2.3 Jordan's boxplots

From the left-side graph of Fig. 2.3, we see that the number of calls ranges from 5 to 25, and the median value is approximately 13.5. Furthermore, the interquartile range (IQR) is given by Eq. (2.9):

$$IQR = Q_3 - Q_1$$

$$= 17.25 - 9 = 8.25$$

since Q_1 and Q_3 are 17.25 and 9, respectively.

From the right-side graph of Fig. 2.3, we see that the sales range from 4 to 21, and the median value is approximately 12. Furthermore, the interquartile range (IQR) is given by Eq. (2.9):

$$IQR = Q_3 - Q_1$$

$$= 14.75 - 7 = 7.75$$

since Q_1 and Q_3 are 14.75 and 7, respectively.

Notice that the y-axis differs minimally enough that the boxplots look relatively similar despite the different values of the quartiles, minimum, and maximum.

(f) Using the boxplot function, we directly plug in the dataframe df.

```
boxplot(df)
```

While the interpretation is similar to part (e), in Fig. 2.4, we can visualize the difference between sales and calls. Clearly, the number of calls is higher for each quartile, the minimum, and the maximum.

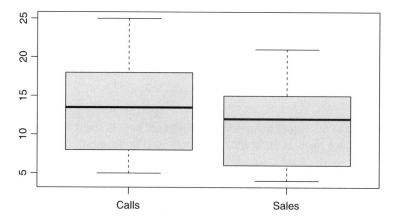

Fig. 2.4 Jordan's boxplot

2.10   Case Study: Top Companies

2.10.1   Problem Statement

One crucial part of business analytics is creating informative visualizations since this area is necessary to understand data across different business domains. Understanding data can lead to actionable results in business. In this case study, we present a dataset consisting of 200 of the world's top companies and provide a quick analysis to help understand datasets.

To understand the overall global economy, we can observe the world's largest public companies using four metrics: assets, market value, sales, and profits. The dataset for the year 2022 consists of 6 columns and 200 rows. The columns in the data are:

- ID—numeric integer corresponding to the row number
- Country—the country in which the company resides
- Sales—the annual sales in billions of USD
- Profit—the annual profit in billions of USD
- Assets—the total company assets in billions of USD
- Value—the company's total market value in billions of USD

As an analyst working for an investment firm, you are tasked with analyzing the top 200 global companies from 2022 to get some understanding of the overall economy. The tasks are the following:

1. Describe the data by running some R commands to familiarize yourself with the data.
2. Generate scatterplots of all numeric variables, particularly sales and profits.
3. Generate histograms of all numeric variables except the ID.

After describing the data and generating the plots, we must interpret the resulting graphs.

2.10.2 Data Description

The first step in analyzing a dataset is to get familiar with the data set of interest. After loading the data, examine the first six observations using the head function.

```
df = read.csv("Top200.csv")
head(df)
```

```
##   ID Country  Sales Profit  Assets   Value
## 1  1      US 15.264  6.261 234.234  55.157
## 2  2      US 31.454 11.662  12.398   6.204
## 3  3   China 36.713 16.065  46.544 184.417
## 4  4   Japan  0.743 -6.856   0.765  29.102
## 5  5      US  2.511  7.644  26.513  48.270
## 6  6   Japan 28.391  4.222 258.418  40.879
```

Note that half of the companies listed in the head output are from the United States, and one-third are from Japan. We now use the str (structure) command to view the dataframe df.

```
str(df)
```

```
## 'data.frame':    200 obs. of  6 variables:
##  $ ID     : int  1 2 3 4 5 6 7 8 9 10 ...
##  $ Country: chr  "US" "US" "China" "Japan" ...
##  $ Sales  : num  15.264 31.454 36.713 0.743 2.511 ...
##  $ Profit : num  6.26 11.66 16.07 -6.86 7.64 ...
##  $ Assets : num  234.234 12.398 46.544 0.765 26.513 ...
##  $ Value  : num  55.2 6.2 184.4 29.1 48.3 ...
```

From the structure, we note the variable type of each variable within the dataframe. Specifically, ID is an int or integer, Country is a chr or character string, and the remaining variables are num or numeric.

The summary function provides insight into the data by displaying some summary statistics.

```
summary(df)
```

```
##       ID            Country              Sales
##  Min.   : 1.00   Length:200         Min.   : 0.034
```

```
##   1st Qu.: 50.75    Class :character    1st Qu.:   6.572
##   Median :100.50    Mode  :character    Median :  14.932
##   Mean    :100.50                       Mean    :  19.820
##   3rd Qu.:150.25                        3rd Qu.:  27.538
##   Max.    :200.00                       Max.    :126.626
##        Profit           Assets              Value
##   Min.    :-9.619   Min.    :  0.351   Min.    :  0.131
##   1st Qu.: 1.260    1st Qu.: 26.328    1st Qu.: 12.813
##   Median : 6.058    Median : 64.045    Median : 31.363
##   Mean    : 6.296   Mean    : 87.662   Mean    : 44.728
##   3rd Qu.:10.380    3rd Qu.:123.857    3rd Qu.: 58.949
##   Max.    :28.638   Max.    :507.224   Max.    :263.381
```

The output shows the integer and numeric columns' minimum, first quartile, median, mean, third quartile, and maximum values. From observing this data, note the minimum values are at or near 0 for Sales, Assets, and Value. On the other hand, Profit has a negative minimum value. Further note that the third quartiles are much different than the maximum values for these numeric variables, which denotes a trailing tail in each distribution. Since the third quartile is the 75th percentile, we employ the quantile function to verify the third quartile from the summary. The first argument in the quantile is the vector we access from df using the $ for each numeric variable.

```
Q3Sales = quantile(df$Sales, .75)
Q3Profit = quantile(df$Profit, .75)
Q3Assets = quantile(df$Assets, .75)
Q3Value = quantile(df$Value, .75)
```

The variable values can be seen within the Posit environment or by typing the variable name in the console. For demonstration, we print Q3Sales by typing the R code.

```
Q3Sales
```

```
##       75%
## 27.53775
```

Notice that 27.53775 is shown as the third quartile for Sales, which is consistent with the summary.

2.10.3 Scatterplots

To generate a plot of Sales and Profit, the plot function is executed with df$Sales as the first argument for the x values, and df$Profit is the second argument to denote the corresponding y values.

```
plot(df$Sales, df$Profit)
```

While this plot is correct, it chooses the *x*- and *y*-labels as df$Sales and df$Profit, respectively. Rather than use these undesirable variable names, we can use the following notation to plot the data. In this case, Profit precedes Sales and is separated by the tilde character ("~"). Further note, the dataframe does not precede the variable since the data were specified by data = df.

```
plot(Profit ~ Sales, data = df)
```

Both previous lines of code will produce a plot with Sales on the *x*-axis and Profit on the *y*-axis. While the second plot has slightly more appealing labels, either can be modified to particularly specify the axis labels using the xlab and ylab arguments. In addition, we can add a title using the main option.

```
plot(df$Sales, df$Profit, xlab = "Sales (Billions $)",
     ylab="Profits (Billions $)",
     main = "Scatterplot of the Top 200 Companies")
```

From the plot in Fig. 2.5, each circle represents an observation in the data. These unfilled circles display the density of the observations. For example, whenever an observation is near another similar observation, the overlap is easily seen.

Next, we can generate a scatterplot matrix of all numeric variables (excluding ID). The numeric variables are 3, 4, 5, and 6 within the dataframe. Here we reference these variables in the dataframe with the c function.

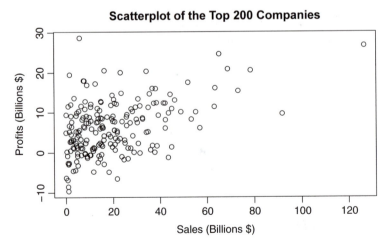

Fig. 2.5 Scatterplot of sales and profits

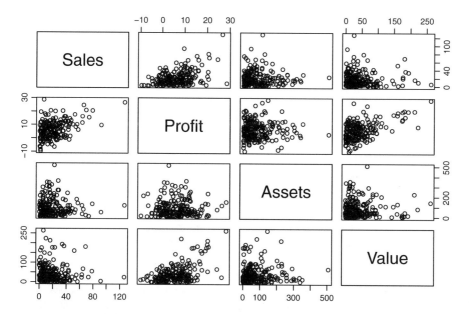

Fig. 2.6 Scatterplot matrix of the top 200 companies

```
numeric_df = df[,c(3,4,5,6)]
```

The resulting `numeric_df` is a dataframe consisting of the numeric columns. Passing `numeric_df` as the only argument into the plot function produces a matrix of scatterplots.

```
plot(numeric_df)
```

Along the diagonal of Fig. 2.6, we see the variable names. Notice all scatterplots in the top row have `Sales` on the *y*-axis, and all scatterplots in the first column have `Sales` on the *x*-axis. The scatterplot in the first row and second column shows `Profit` versus `Sales`. A matrix of scatterplots can help assess the relationship between variables quickly. Remember that this visualization becomes less appealing when too many variables are included since the graphs will become smaller and less visible. While the scatterplots are not extremely revealing, it can be seen that `Profit` does tend to increase with `Sales` and `Value`.

2.10.4   Histograms

Using the `hist` function, we generate a histogram of `Sales` by passing the `df$Sales`.

```
hist(df$Sales)
```

The plot generated from the previous line of code will have suboptimal labels. Therefore, we carefully specify the xlab and main arguments in the hist function to generate histograms for Sales, Profit, Assets, and Value. We employ the par function here, which allows multiple plots to be shown by specifying the mfrow argument. Since the mfrow option is c(2,2), there will be two rows and two columns of plots.

```
par(mfrow=c(2,2))
hist(df$Sales, xlab = "Sales (Billions $)", main = "Sales")
hist(df$Profit, xlab = "Profit (Billions $)", main = "Profit")
hist(df$Assets, xlab = "Assets (Billions $)", main = "Assets")
hist(df$Value, xlab = "Market Value (Billions $)",
     main = "Market Value")
```

As mentioned in the interpretation of the summary statistics, all numeric variables have a tail leading to the right, referred to as a right skew. We see from Fig. 2.7 that a bulk of sales is below 60 billion USD, a majority of assets are below 150 billion USD, and a majority of market values are below 100 billion USD. Profit has most values between 0 and 20 billion, but there are also many companies with negative profits.

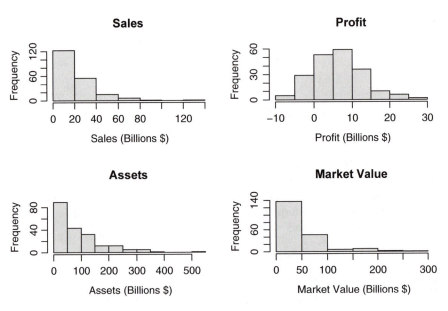

Fig. 2.7 Histograms of the top 200 companies

2.10.5   Case Conclusion

In this analysis, we showed the value of visualizations by creating some elementary plots characterizing the top 200 dataset. In addition, we interpreted the data by using some essential R functions. Analyzing data from visualizations and basic R functions is a valuable skill that analysts need to conclude their results.

Problems

1. **Accounting Basic Statistics**

   Using Table 2.3, answer the questions below. Both variables are in thousands of U.S. dollars.

   Without the use of a computer:

   a. Calculate the mean for both X and Y.
   b. Calculate the standard deviation for both X and Y.

2. **Sales Calls Basic Statistics**

   Jordan's colleague at Enright Associates is Abdullah who also manages retirement accounts. Jordan documented the number of sales calls Abdullah made each day for the last week. The number of calls is denoted by X, and the number of sales that resulted is denoted as Y. The values of X and Y are given in Table 2.4. Without the use of a computer:

   a. Calculate the mean for both X and Y.
   b. Calculate the standard deviation for both X and Y.
   c. Compute the variance for both X and Y.

Table 2.3 Basic accounting data

Adjusted gross income (X)	Itemized deductions (Y)
51	11
142	31
66	11
67	14
35	12

Table 2.4 Abdullah's calls and sales

Variable							
Calls (X)	18	23	25	15	8	20	12
Sales (Y)	10	11	15	8	3	11	7

3. **Sales Calls Basic Statistics in R**

Investigate the relationship between X and Y from the previous problem by doing the following. While using a computer (with R):

   a. Create vectors for X and Y.
   b. Calculate the mean for both X and Y.
   c. Calculate the standard deviation for both X and Y.
   d. Compute the variance for both X and Y.

4. **Sales Calls Plots in R**

Investigate the relationship between X and Y from the previous problem by doing the following. While using a computer (with R):

   a. Create a dataframe containing calls and sales.
   b. Use the `plot` function to create a scatterplot of calls and sales. Interpret the plot.
   c. Use the `hist` function to create histograms of both calls and sales. Interpret the plots.
   d. Use the `boxplot` function to create a boxplot of calls and sales on separate plots. Interpret the plots.
   e. Use the `boxplot` function to create a boxplot of calls and sales on the same plot. Interpret the plot.

5. **Top 200 Companies Basic Statistics**

In the chapter case study, the summary function was used to provide summary statistics of the variables. Use the Top 200 dataset to answer the following questions:

   a. Calculate the mean of each numeric variable.
   b. Calculate the standard deviation of each numeric variable.
   c. Find the sum of each numeric variable.

Chapter 3
Regression Fundamentals

There is bound to be a regression toward the mean.

—Charlie Munger

3.1 Introduction

With a recent expansion of information collection and storage, businesses increasingly turn to classical analyses of data. In particular, linear regression analysis, while developed more than 200 years ago, remains a fundamental concept in statistics and business analytics. Linear regression is at the heart of many predictive methods, including modern machine learning models.

In this chapter, we examine the core concepts of linear regression. The discussion of linear regression will be limited to two variables to focus on developing a clear understanding of the calculations and theory. The concepts are put into context with a simple application using data from a call center, predicting successful sales by the number of calls.

In the final discussion of the chapter, we present and solve an accounting case study involving the prediction of the adjusted gross income of an individual based on the amount of their itemized deductions. This case study makes use of the R programming language with descriptive graphs and the relevant source code.

3.2 Covariance

When analyzing a single random variable, the mean and standard deviation are often the only statistics needed since they represent the center and spread of the variable's distribution. When two random variables are analyzed for a relationship,

© The Author(s), under exclusive license to Springer Nature Switzerland AG 2023
D. P. McGibney, *Applied Linear Regression for Business Analytics with R*,
International Series in Operations Research & Management Science 337,
https://doi.org/10.1007/978-3-031-21480-6_3

the covariance can be calculated. For random variables X and Y with data pairs (x_i, y_i), the joint sum of squares of X and Y can be calculated as

$$SS_{xy} = \sum_{i=1}^{n}(x_i - \bar{x})(y_i - \bar{y}) \tag{3.1}$$

or equivalently as

$$SS_{xy} = \sum_{i=1}^{n} x_i y_i - \frac{\left(\sum_{i=1}^{n} x_i\right)\left(\sum_{i=1}^{n} y_i\right)}{n}. \tag{3.2}$$

The covariance can be calculated from the sum of squares of X and Y as

$$Cov(X, Y) = \frac{SS_{xy}}{n - 1}. \tag{3.3}$$

3.3   Correlation Coefficient

To predict an unknown variable, knowing which variables are highly correlated with the unknown variable of interest proves helpful. For example, knowing that height and weight are correlated is useful if you would like to approximate the weight of an individual based on their height. It is also useful if the degree of correlation between two random variables can be measured. The Pearson correlation coefficient, referred to as the correlation coefficient herein, provides such a measurement. For height and weight values, the correlation coefficient would be relatively high since there is a relationship between the two variables.

Similar to the covariance, the correlation coefficient describes the relationship between two or more random variables. The correlation coefficient differs from the covariance because the correlation is scaled and therefore unitless.

The sample correlation coefficient can be calculated using the equation:

$$r = \frac{Cov(X, Y)}{s_x s_y}, \tag{3.4}$$

where s_x and s_y are the sample standard deviations of X and Y, respectively. The sample correlation coefficient can also be solved in terms of the sum of squares:

$$r = \frac{SS_{xy}}{\sqrt{SS_{xx} SS_{yy}}}. \tag{3.5}$$

The correlation coefficient has several properties:

1. It is a unitless measurement between -1 and 1 inclusive ($-1 \leq r \leq 1$).
2. If $r = 1$, there is perfect positive linear correlation. If $r = -1$, there is perfect negative linear correlation. If $r = 0$, there is no linear correlation.
3. Positive values of r imply that as x increases, y tends to increase. Negative values of r imply that as x increases, y tends to decrease.
4. The value of r is the same for the pairs (x, y) and the corresponding pairs (y, x).
5. The value of r does not change when either variable is converted into different units.
6. The correlation coefficient only measures the degree of linear correlation.

While the correlation coefficient is very useful in finding a measure of correlation between variables, it is limited since it only measures the degree of linear correlation between variables. Some may wrongfully assume that if there is a high correlation coefficient, then the relationship must be linear; however, there are many relationships that are nonlinear but close to linear. Another misinterpretation of the correlation coefficient occurs in the case of a low value of r. If the value of r is 0, it does not mean there is no relation between X and Y, but rather that there is no *linear* relationship between X and Y. While this may seem counter-intuitive, consider the example in Fig. 3.1 where a clear relationship between X and Y exists. Because the relationship is circular, the correlation coefficient is 0.

Fig. 3.1 Nonlinear correlation

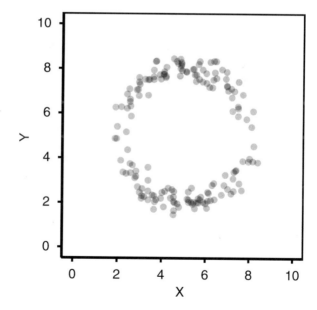

3.4   Coefficient of Determination

The square of the correlation coefficient (r) is called the coefficient of determination but is more commonly referred to as R^2. The value of R^2 is obtained by squaring Eq. (3.5):

$$R^2 = \left(\frac{SS_{xy}}{\sqrt{SS_{xx} SS_{yy}}} \right)^2$$

$$R^2 = \frac{SS_{xy}^2}{SS_{xx} SS_{yy}}. \tag{3.6}$$

The R^2 value signifies an important measure of how well the model fits the data since it measures the amount of variability in Y that is explained by the variability in X. For convenience and interpretability, R^2 is often converted into a percentage and represents the percentage of the explained variability in Y.

If R^2 represents the fraction of explained variability, it follows that the value $1 - R^2$ is the ratio of *unexplained* variability in Y. This unexplained variation can be the result of random chance and/or lurking variables. Obviously, one desires to maximize R^2 and thus minimize $1 - R^2$.

Many students inquire: what constitutes a high R^2 or a high r value? While there is no specific threshold that separates high from low correlation, one can get a general idea about what constitutes a high correlation and a low one through careful observation. Referring to Fig. 3.2, the first scatterplot at the top-left of the page shows a perfect positive linear correlation that has an r value of 1. The next three scatterplots on the left represent r values of 0.8, 0.5, and 0.3. The last scatterplot at the bottom of the figure corresponds to a r value of 0. Note that the data trend is still discernible at $r = 0.8$, but the trend slowly degrades as r decreases. In the same figure, in the upper right, the values of r are -1, -0.8, -0.5, and -0.3. From these plots, it should be evident the magnitude of r denotes the degree of linear relationship, whereas the sign denotes whether the linear trend is increasing or decreasing. The sign of the slope for a linear regression is always in agreement with the slope of the r value. The value of R^2 does not take this upward or downward trend into account as seen in the figure (Fig. 3.2); particularly, the spread around the line is relatively similar for $r = 0.8$ and $r = -0.8$.

3.5   Sales Calls Application: Variable Correlations

Here, we investigate Jordan's data further. Using Table 2.2 from the previous chapter, investigate the relationship between X and Y by doing the following:

(a) Calculate the sum of squares of X and Y.

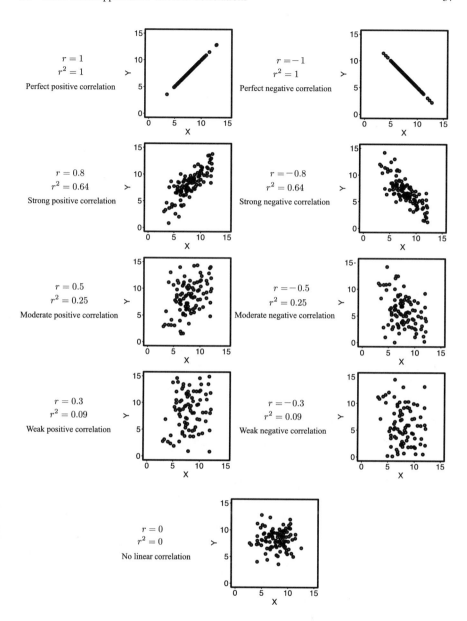

Fig. 3.2 Correlation values

(b) Calculate the sample covariance.

(c) Compute the sample correlation coefficient.

(d) Compute the coefficient of determination.

(e) Plot a scatterplot.

Solution

(a) First, calculate the following values required to obtain SS_{xy}:

$$\sum_{i=1}^{n} x_i = 12 + 18 + 5 + 25 + 15 + 8 = 83$$

$$\sum_{i=1}^{n} y_i = 10 + 15 + 4 + 21 + 14 + 6 = 70$$

$$\sum_{i=1}^{n} x_i y_i = 12 \times 10 + 18 \times 15 + 5 \times 4 + 25 \times 21 + 15 \times 14 + 8 \times 6 = 1193.$$

Knowing these three values, Eq. (3.2) can now be utilized:

$$SS_{xy} = \sum_{i=1}^{n} x_i y_i - \frac{\left(\sum_{i=1}^{n} x_i\right)\left(\sum_{i=1}^{n} y_i\right)}{n}$$

$$SS_{xy} = 1193 - \frac{(83)\,(70)}{6} = 224.67.$$

(b) The sample covariance can be calculated from Eq. (3.3)

$$Cov(X, Y) = \frac{SS_{xy}}{n - 1}$$

$$Cov(X, Y) = \frac{224.67}{5} = 44.93.$$

(c) To calculate the sample correlation coefficient, one must first calculate the sum of squares for Y:

$$SS_{yy} = \sum_{i=1}^{n} y_i^2 - \left(\sum_{i=1}^{n} y_i\right)^2 / n$$

$$SS_{yy} = 1014 - (70)^2 / 6 = 197.33.$$

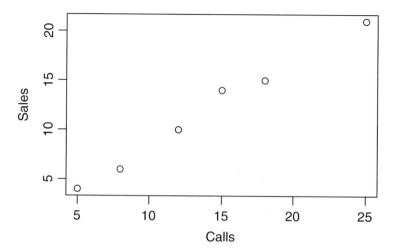

Fig. 3.3 Scatterplot of sales vs. calls

From earlier calculations, recall $SS_{xx} = 258.83$ and $SS_{xy} = 224.67$. The correlation coefficient is calculated from Eq. (3.5),

$$r = \frac{SS_{xy}}{\sqrt{SS_{xx} SS_{yy}}}$$

$$r = \frac{224.67}{\sqrt{258.83 \times 197.33}} = 0.994.$$

(d) The coefficient of determination can be found by squaring the value of r:

$$R^2 = 0.994^2 = 0.988$$

indicating that 98.8% of the variation in Y (sales) can be explained by the variation in X (calls).

(e) The scatterplot is shown in Fig. 3.3.

3.6   Least Squares Criterion

The least squares criterion is a method that determines the best line between two variables. The best line is achieved by making the distance between the line and the points as small as possible. This is depicted in Fig. 3.4. These vertical distances are known as residuals, and the ith residual is expressed as

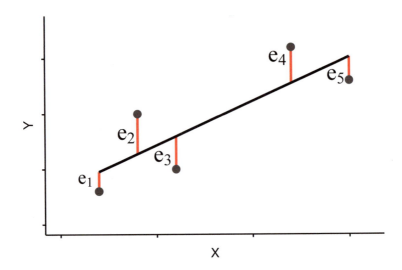

Fig. 3.4 Least squares depiction

$$e_i = y_i - \hat{y}_i. \tag{3.7}$$

Summing the residuals poses a problem mathematically since some residuals are positive, while others are negative. It is therefore desirable to minimize the magnitude of these distances, which can be achieved by minimizing the sum of the squares of the vertical distances from the data points (x_i, y_i) to the line that is referred to as the sum of the squared error (SSE), expressed as

$$SSE = \sum_{i=1}^{n} e_i^2 \tag{3.8}$$

$$SSE = \sum_{i=1}^{n} (y_i - \hat{y}_i)^2. \tag{3.9}$$

Using the least squares criterion, one can find the equation of a line, which can be referred to as the regression line. The equation of the regression line is simply called the regression equation and is given by

$$\hat{Y} = \hat{\beta}_0 + \hat{\beta}_1 X, \tag{3.10}$$

where:

- $\hat{Y}$ denotes an estimate of Y.
- $\hat{\beta}_0$ and $\hat{\beta}_1$ estimate the Y-intercept and the slope, respectively.

The values of $\hat{\beta}_0$ and $\hat{\beta}_1$ are solved for by finding the minimum value of Eq. (3.18) while substituting in Eq. (3.10) for $\hat{y}$:

$$SSE = \sum_{i=1}^{n} (y_i - (\hat{\beta}_0 + \hat{\beta}_1 x_i))^2$$

$$SSE = \sum_{i=1}^{n} (y_i - \hat{\beta}_0 - \hat{\beta}_1 x_i)^2. \qquad (3.11)$$

To minimize Eq. (3.11) and thereby find equations for $\hat{\beta}_0$ and $\hat{\beta}_1$, the partial derivative of the SSE with respect to $\hat{\beta}_0$ and $\hat{\beta}_1$ should be found and set equal to 0 as is common practice in calculus. The resulting equations are referred to as the "normal" equations.

The result of solving the normal equations is

$$\hat{\beta}_1 = \frac{SS_{xy}}{SS_{xx}} \qquad (3.12)$$

$$\hat{\beta}_0 = \bar{y} - \hat{\beta}_1 \bar{x}, \qquad (3.13)$$

where $\bar{x}$ and $\bar{y}$ denote the sample means of X and Y, respectively.

The regression line has the following properties:

1. The point $(\bar{x}, \bar{y})$ is always on the linear regression line.
2. The value $\hat{Y}$ is read "Y hat," where the "hat" denotes an estimate.
3. The variable X is referred to as the independent variable, predictor, or explanatory variable.
4. The variable Y is referred to as the dependent variable, response, or target variable.
5. The values of X can be plugged into the regression equation to get estimates of Y denoted as $\hat{Y}$.

With the knowledge to properly solve the regression equation, we are equipped with the necessary tools to continue the sales calls application. Previously, we identified a positive linear correlation between calls and sales that suggested further analysis.

3.7 Sales Calls Application: Simple Regression

From the data used in the previous application (Table 2.2), the least squares regression line can be computed. Further investigate the relationship between X and Y by doing the following:

(a) Find the regression equation that best fits the data.
(b) Interpret the slope of the regression equation.

Solution

(a) Recall from the previous problems that $SS_{xx} = 258.83$ and $SS_{xy} = 224.67$. Using these values and Eq. (3.12), the slope is

$$\hat{\beta}_1 = \frac{SS_{xy}}{SS_{xx}}$$

$$\hat{\beta}_1 = \frac{224.667}{258.833} = 0.8680.$$

Using the slope and the sample means ($\bar{x} = 13.8333$ and $\bar{y} = 11.6667$), the intercept is given from Eq. (3.13):

$$\hat{\beta}_0 = \bar{y} - \hat{\beta}_1 \bar{x}$$

$$\hat{\beta}_0 = 11.6667 - 0.868 \times 13.8333 = -0.3406.$$

Note that as a final step, the intercept and slope are plugged into Eq. (3.10)

$$\hat{Y} = \hat{\beta}_0 + \hat{\beta}_1 X$$

$$\hat{Y} = -0.3406 + 0.868X.$$

(b) Since the slope is 0.868, if the number of calls (X) is increased by 1, then the number of sales (Y) is expected to increase by 0.868. This interpretation can be expanded to 10 calls; that is, if the number of calls increases by 10, then the number of sales is expected to increase by 8.68.

3.8   Interpolation and Extrapolation

Predicting a response variable, Y, from a predictor variable value, X, that is within the range of observed X values in the data set is called interpolation. If the value of x_i is outside of the range of observed X values, then the resulting prediction is called an extrapolation.

Since interpolation occurs in the range of known X values, the result is most likely easy to predict, as opposed to the case when x_i is far away from any similar known X values. Extrapolation should hence be met with skepticism since the available data are not similar enough to the X value of interest. Furthermore, it is often the case that patterns appear to be linear, but when extreme points are considered, a nonlinear trend emerges, which is why caution should be taken when

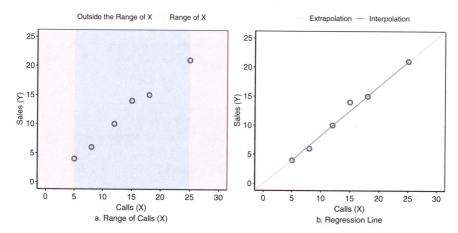

Fig. 3.5 Interpolation and extrapolation

extrapolating. For instance, if Jordan makes 100 calls in a day, she probably will not be able to maintain her typical sales conversion rate than if she made 10 calls in a day.

In Fig. 3.5a, the minimum value of X can be deciphered as 5. To the left of $X = 5$, there are no data values, which means that the interval of $X < 5$ is outside the range of X. The area to the left of $X = 5$ is therefore denoted in pink (in Fig. 3.5a) since it is outside the range of X. Similarly, the maximum value of X is 25, and therefore, the area to the right of $X = 25$ is in pink.

Using the range information from Fig. 3.5a along with the regression line, Fig. 3.5b shows which portions of the regression line are extrapolation and which are interpolation. As mentioned previously, the extrapolated values are values predicted with X values outside of the range of X.

3.9 Sales Calls Application: Prediction

Jordan is considering a few ways to increase her sales numbers including making 3 calls if she is sick, working a half day on Saturday when she could make 5 calls, increasing her minimum number of calls to 15 per weekday, and increasing the number of calls she makes to 30 on Mondays. Using Table 2.2, investigate the relationship between X and Y by doing the following:

(a) Find the range of X.
(b) Predict the corresponding sales for $x = 3, 5, 15, 30$.
(c) Indicate whether each prediction from (b) is an interpolation or extrapolation.

Solution

(a) The smallest value of X is 5 and the largest is 25. Therefore, the range is between 5 and 25 inclusive:

$$5 \le X \le 25.$$

Figure 3.5a depicts the range of X for this application.

(b) The least squares regression line was computed in the previous application:

$$\hat{Y} = -0.3406 + 0.868X.$$

Using the regression line, the predictions are made as follows:

- When $x = 3$, then $\hat{y} = -0.3406 + 0.868(3) = 2.2634$.
- When $x = 5$, then $\hat{y} = -0.3406 + 0.868(5) = 3.9994$.
- When $x = 15$, then $\hat{y} = -0.3406 + 0.868(15) = 12.6794$.
- When $x = 30$, then $\hat{y} = -0.3406 + 0.868(30) = 25.6994$.

(c) Since $x = 3$ and $x = 30$ are outside the range of X indicated in part a, the corresponding predictions are extrapolations. The values of $x = 5$ and $x = 15$ are within the range of X and therefore are interpolations.

3.10   Explained Deviation

To understand how R^2 can measure the percentage of explained variation, one should consider the deviations that make up the variation. Figure 3.6 depicts the explained, the unexplained, and the total deviations for one particular observed value (x, y) that is above the regression line. From the depiction, the total deviation from the mean of Y ($\bar{y}$) is $y - \bar{y}$. This deviation, or difference, consists of both explained and unexplained deviation. For the observed value of x, the linear regression prediction of y would be $\hat{y}$. Therefore, the amount of deviation explained by linear regression would be $\hat{y} - \bar{y}$. The amount of unexplained deviation is $y - \hat{y}$. Since the total deviation can be broken down into unexplained and explained deviations, the deviations can be expressed for the particular value of Y as

$$(y - \bar{y}) = (\hat{y} - \bar{y}) + (y - \hat{y}), \tag{3.14}$$

or for the entire vector Y as

$$(Y - \bar{y}) = (\hat{Y} - \bar{y}) + (Y - \hat{Y}). \tag{3.15}$$

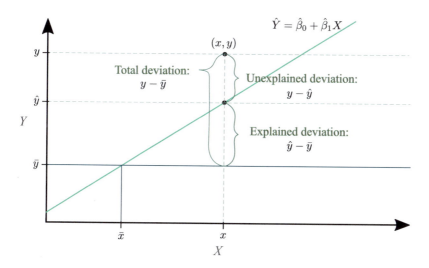

Fig. 3.6 Explained vs. unexplained deviation

The deviation can be found from a single observation, but the amount of variation is calculated using every observation. By squaring and summing the deviations that comprise the total deviation of each observation, a measure of the total variation is

$$SST = \sum_{i=1}^{n}(y_i - \bar{y})^2,$$ (3.16)

which is referred to as the sum of squares total (SST). The variation of the sum of squares due to regression (SSR) is

$$SSR = \sum_{i=1}^{n}(\hat{y}_i - \bar{y})^2,$$ (3.17)

and the variation of the sum of squares due to error (SSE) is

$$SSE = \sum_{i=1}^{n}(y_i - \hat{y}_i)^2.$$ (3.18)

Similar to Eq. (3.15), the SST can be expressed in terms of the SSR and SSE:

$$SST = SSR + SSE$$ (3.19)

or equivalently stated,

$$\sum_{i=1}^{n}(y_i - \bar{y})^2 = \sum_{i=1}^{n}(\hat{y}_i - \bar{y})^2 + \sum_{i=1}^{n}(y_i - \hat{y}_i)^2. \qquad (3.20)$$

As mentioned previously, the R^2 value is the amount of explained variation. Therefore, using SSR and SST, the value of R^2 is

$$R^2 = SSR/SST. \qquad (3.21)$$

Equation (3.21) is mathematically equivalent to Eq. (3.6), which is left for the reader to prove as one of the problems at the end of the chapter.

In the next sections, concepts from this chapter are brought together to solve a case study in accounting analytics.

3.11   Case Study: Accounting Analytics

3.11.1   Problem Statement

Prediction and analytics constitute valuable skills to have in many different business domains. For this reason, the skills required for entering the job market have evolved, with analytics now being one of the most widely sought-after skill sets. Analytics has especially changed the field of accounting, in which irregularities in the data can be quickly found using simple analytical methods.

As an accounting analyst, you are investigating the amount of itemized deductions that are reasonable considering someone's adjusted gross income. You are handed the tax return documents of 8 individuals, and you are tasked with analyzing these data. From these data, you are to generate a scatterplot, calculate the correlation coefficient, fit a linear regression model, and use your fitted model to predict the itemized deductions for an individual who is at risk of an audit. The data are listed in Table 3.1 where both variables are in thousands of U.S. dollars.

Table 3.1 Accounting data

Adjusted gross income	Itemized deductions
77	16
96	15
50	8
54	10
130	21
67	13
135	24
114	24

3.11.2   *Correlation and Scatterplot*

From the data in Table 3.1, the variable that will be predicted is the amount of itemized deductions. Therefore, the itemized deductions variable is the response variable, which will be assigned as Y by convention. The variable that will be used to predict Y is known as the predictor variable and is assigned as X by convention. Using the c function in R, vectors are created for both x and y.

```
x = c(77, 96, 50, 54, 130, 67, 135, 114)
y = c(16, 15, 8, 10, 21, 13, 24, 24)
```

As an initial action, the points can be plotted (as shown in Fig. 3.7) using the plot command:

```
plot(x, y, xlab = "Adj. Gross Income (in $1000s)",
     ylab = "Itemized Deductions (in $1000s)")
```

The relationship can be observed in the plot. However, it is beneficial to quantify the relationship using the correlation coefficient (r) and also the coefficient of determination (R^2). Using the cor function allows for the calculation of the correlation coefficient:

```
cor(x,y)
```

```
## [1] 0.9426905
```

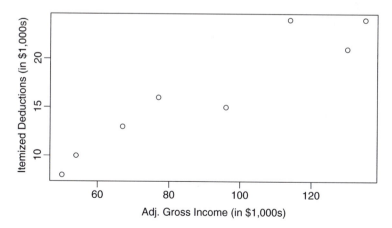

Fig. 3.7 Accounting data scatterplot

The value of 0.9427 denotes a strong positive linear correlation between the response and the predictor variable. The coefficient of determination (R^2) can be found by squaring the correlation coefficient:

```
cor(x,y)^2
```

```
## [1] 0.8886653
```

In interpreting the model, one should note that the R^2 value is relatively high at 0.8887. From this value, one can conclude that 88.87% of the variation in Y can be explained by the variation in X. More specifically, the variation in the adjusted gross income can explain 88.87% of the variation in the itemized deductions.

3.11.3   Linear Regression Modeling

One can further analyze the relationship between adjusted gross income and itemized deductions using linear regression. If a linear relationship exists between X and Y, or even if the relationship is approximately linear, the simple linear regression model can be used. The coefficients of the simple linear regression model can be calculated using the lm function. Assigning reg to be the lm object is done using:

```
reg = lm(y ~ x)
```

A summary of the linear regression can be obtained using the summary function:

```
summary(reg)
```

```
##
## Call:
## lm(formula = y ~ x)
##
## Residuals:
##     Min      1Q  Median      3Q     Max
## -2.3448 -1.6123 -0.0862  0.9739  3.5520
##
## Coefficients:
##             Estimate Std. Error t value Pr(>|t|)
## (Intercept)  0.79430    2.38250   0.333 0.750184
## x            0.17240    0.02491   6.920 0.000451 ***
## ---
## Signif. codes:  0 '***' 0.001 '**' 0.01 '*' 0.05 '.' 0.1 ' ' 1
##
## Residual standard error: 2.204 on 6 degrees of freedom
## Multiple R-squared:  0.8887, Adjusted R-squared:  0.8701
## F-statistic: 47.89 on 1 and 6 DF,  p-value: 0.0004506
```

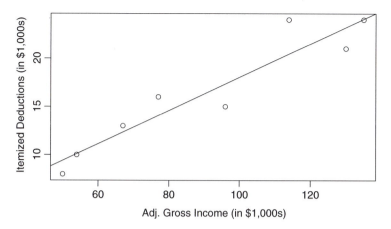

Fig. 3.8 Accounting data scatterplot

From the summary output of the regression line, the coefficient estimates are given in the "Estimate" column. From this column, the estimate of the intercept is displayed as 0.79430, and the estimate of the slope is displayed as 0.17240. The resulting regression equation from these coefficients is

$$\hat{Y} = 0.79430 + 0.17240X.$$

As a further note from the summary, the "Multiple R-squared" value matches that of the R^2 value from squaring the r value found using the cor function. The value of R^2 is referred to as Multiple R^2 when there are multiple predictor variables. Since there is only one predictor variable in our model, the "Multiple R-squared" designation is inappropriate in this case.

The scatterplot from Fig. 3.7 can be modified to include the linear regression line within the plot using the abline function as shown in Fig. 3.8. The abline function should be used following the plot command and only requires one input: the regression object created by the lm function.

```
plot(x, y, xlab = "Adj. Gross Income (in $1000s)",
     ylab = "Itemized Deductions (in $1000s)")
abline(reg)
```

3.11.4 Audit Scenarios

After analyzing the data, your manager identifies 4 individuals who will most likely be audited due to irregularities in their income data and the amount of itemized deductions that they have reported. The adjusted gross incomes from last year for the individuals are:

1. $30,000
2. $78,000
3. $214,000
4. $120,000

From the model, an appropriate number of itemized deductions can be predicted with the predict function. The values will need to be combined into a dataframe in order to pass them through the predict function. A dataframe can be created using the data.frame function:

```
df = data.frame(x = c(30, 78, 214, 120))
```

After the adjusted gross income values are specified within the new dataframe, the predict function can be used to approximate the amount of itemized deductions using the fitted model (reg). The predict function requires the fitted model object as the first argument and the new dataframe as the second argument.

```
predict(reg, df)
```

```
##         1         2         3         4
## 5.966315 14.241543 37.688022 21.482367
```

Note the third value in df is $214,000, which is outside the range of X, and therefore means the third prediction (37.688022 or $37,688) represents an extrapolation.

After reporting your predictions to your manager, she notes that the reported itemized deductions for each individual are:

1. $6849
2. $12,862
3. $24,690
4. $42,409

The results indicate the first and second individuals are roughly close to the predicted amount of itemized deductions. The third individual corresponds to the extrapolation, which has a large difference between the observed value and the predicted value. However, more data should be gathered before assuming the third individual's itemized deductions are irregular. The fourth individual has approximately twice the predicted itemized deductions and should be scrutinized. While there are many more intricacies to tax and audit analysis, this case study shows the value that predictive analytics can play in the area of accounting.

3.11.5   Case Conclusion

In this analysis, we showed the value of analytics by fitting a model to predict the amount of itemized deductions using the adjusted gross income as a predictor. The tax code has many more significant numeric variables that one can consider for a detailed analysis. For this reason, accounting professionals have sought analytical skills as a means to have an advantage over their competition. Analytics will continue to play an important role in accounting and accounting-related applications.

Simple linear regression—regression with one predictor variable—is fundamental in understanding regression analysis. This case study demonstrates the utility of that analysis, using some fundamental R commands that were used, such as: c, lm, abline, summary, cor, predict, plot, and data.frame. When studying more complex regression analysis, several of these commands will need to be used further.

Problems

1. **Accounting Analytics Basic Statistics**
   In the chapter case study, the problem solved involved the use of R programming. The same analysis can be done without the use of a computer as well. Using Table 3.1, answer the questions below. Both variables are in thousands of U.S. dollars.
   Without the use of a computer:

   a. Find the correlation coefficient and the coefficient of determination for X and Y.
   b. Find the equation of the least squares line between X and Y.

2. **Accounting Analytics Prediction**
   From the accounting analytics case study, it was shown that the regression equation was:

$$\hat{Y} = 0.7942958 + 0.1724006X.$$

   Without the use of a computer:

   a. Find the range of X using the values of X from Table 3.1.
   b. Predict the corresponding sales for $x = 30, 55, 115, 300$.
   c. Indicate whether each prediction from part b is an interpolation or extrapolation.

Table 3.2 Social media
influencers data

Pay	Clicks
1000	26,855
250	2230
5000	99,323
800	23,321
2500	47,916
780	3274
100	4686
980	15,591
400	7174
4200	92,447
270	10,219
3600	59,221

3. **Sales Calls Problem**

   Jordan's colleague at Enright Associates is Abdullah who also manages
   retirement accounts. Jordan documented the number of sales calls Abdullah
   made each day for the last week. The number of calls is denoted by X, and the
   amount of sales that resulted is denoted as Y. The values of X and Y are given
   in Table 2.4.

   Without the use of a computer:

   a. Find the correlation coefficient and the coefficient of determination for X
      and Y.
   b. Find the equation of the least squares line between X and Y.
   c. From Jordan's data, a slope of 0.868 was noted. How does Abdullah's slope
      compare with Jordan's?

   While using a computer (with R):

   d. Verify your results from parts a and b.
   e. Generate a summary of the regression line between X and Y.
   f. Plot a scatterplot of X and Y.

4. **Social Media Influencers**

   The CEO of Zen Sports Apparel contracts social media influencers to generate
   clicks to their products. The data shown in Table 3.2 reflect the amount paid (in
   dollars) and the resulting number of clicks.

   Without the use of a computer:

   a. Find the correlation coefficient and the coefficient of determination for the
      pay (X) and clicks (Y).
   b. Find the equation of the least squares line between pay (X) and clicks (Y).
   c. What does the slope coefficient indicate?

Table 3.3 Automotive earnings and price data

Rank	Name	Ticker	EPS	Price
1	Douglas Dynamics	PLOW	0.0021	55.00
2	Volkswagen AG	VWAGY	2.6600	19.29
3	Tesla	TSLA	−4.9200	418.33
4	Ford	F	0.0100	9.30
5	Penske Automotive Group	PAG	5.2800	50.22
6	General Motors	GM	4.6200	36.60

While using a computer (with R):

d. Verify your results from part a.
e. Generate a summary of the regression line between pay (X) and clicks (Y).
f. Plot a scatterplot of pay (X) and clicks (Y).
g. Would it be more beneficial to hire several low-cost social media influencers or hire only one social media influencer with the same amount of money?

5. **Automotive Stocks**
   Table 3.3 displays the data for 6 automotive stocks. The data indicate the earnings per share (EPS) reported on 12/31/2019 and the corresponding stock closing price on the same day.
   Without the use of a computer:

   a. Find the correlation coefficient and the coefficient of determination for the EPS (X) and price (Y).
   b. Find the equation of the least squares line between EPS (X) and price (Y).
   c. Based on the value calculated for r, describe the linear relationship between earnings and price.

   While using a computer (with R):

   d. Verify your results from part a.
   e. Generate a summary of the regression line between EPS (X) and price (Y).
   f. Plot a scatterplot of EPS (X) and price (Y).
   g. Based on the regression equation, calculate the predicted price given an EPS of 2.

6. **Age and Net Worth of Billionaires**
   The age and net worth of 10 billionaires are shown in Table 3.4 where "Age" is in years and "Net Worth" is in billions of U.S. dollars.
   Without the use of a computer:

   a. Find the correlation coefficient and the coefficient of determination for the age (X) and net worth (Y).
   b. Find the equation of the least squares line between age (X) and net worth (Y).
   c. Based on the value calculated for R^2, is there a significant relationship between age and net worth?

Table 3.4 Billionaires data

Rank	Age	Net worth	Source
12	82	1.3	Technology
27	58	2.6	Casinos
61	83	2.7	Software
85	82	4.0	Logistics
100	69	8.6	Hedge funds
123	39	10.3	Metals and energy
192	49	10.4	Business software
383	58	18.7	Video games
763	63	21.4	Consulting
944	99	26.4	Telecom

Table 3.5 Bitcoin data

Month	BTC	SPY
1	0.142	−0.006
2	0.363	0.028
3	0.305	0.042
4	−0.020	0.056
5	−0.354	0.007
6	−0.061	0.019
7	0.188	0.028
8	0.133	0.030
9	−0.072	−0.050
10	0.400	0.074
11	−0.070	−0.008
12	−0.077	0.043

While using a computer (with R):

d. Verify your results from part a.
e. Generate a summary of the regression line between age (X) and net worth (Y).
f. Plot a scatterplot of age (X) and net worth (Y).

7. **Bitcoin and S&P 500**

   The returns of Bitcoin in U.S. Dollars (BTC-USD) are said to relate closely with the overall stock market. Using the returns of the S&P 500 market benchmark (SPY), analyze the returns of Bitcoin with the S&P 500. The data given in Table 3.5 are the monthly returns from 2021.
   Without the use of a computer:

   a. Find the correlation coefficient and the coefficient of determination for the returns of Bitcoin (X) and the S&P 500 (Y).
   b. Find the equation of the least squares line between the returns of Bitcoin (X) and the S&P 500 (Y).

c. Based on the value calculated for r, describe the linear relationship between the returns of Bitcoin (X) and the S&P 500 (Y).

d. Calculate and interpret the coefficient of determination.

While using a computer (with R):

e. Verify your results from part a.

f. Generate a summary of the regression line between the returns of Bitcoin (X) and the S&P 500 (Y).

g. Plot a scatterplot of the returns of Bitcoin (X) and the S&P 500 (Y).

8. **Residual Summation Proof**
   Show that both sides of Eq. (3.15) are equivalent:

   $$(Y - \bar{y}) = (\hat{Y} - \bar{y}) + (Y - \hat{Y}).$$

9. **Sum of Squares of X Proof**
   Show that the equations for SS_{xx} (Eqs. 2.4 and 2.5) are equivalent. In particular, show that:

   $$\sum_{i=1}^{n}(x_i - \bar{x})^2 = \sum_{i=1}^{n} x_i^2 - \left(\sum_{i=1}^{n} x_i\right)^2 / n.$$

10. **Sum of Squares of X and Y Proof**
    Show that the equations for SS_{xy} (Eqs. 3.1 and 3.2) are equivalent. In particular, show that:

    $$\sum_{i=1}^{n}(x_i - \bar{x})(y_i - \bar{y}) = \sum_{i=1}^{n} x_i y_i - \frac{\left(\sum_{i=1}^{n} x_i\right)\left(\sum_{i=1}^{n} y_i\right)}{n}.$$

11. **Normal Equation Proof Part 1**
    Show that the estimated intercept value is given by

    $$\hat{\beta}_0 = \bar{y} - \hat{\beta}_1\bar{x}$$

    by taking the partial derivative of the SSE with respect to $\hat{\beta}_0$ and setting it equal to 0.

12. **Normal Equation Proof Part 2**
    Prove that the partial derivative of the SSE with respect to $\hat{\beta}_1$ results in

    $$\hat{\beta}_1 = \frac{SS_{xy}}{SS_{xx}}.$$

    Hint: You will need to use the result from the previous problem.

Chapter 4
Simple Linear Regression

If you can't explain it simply, you don't understand it well enough.

—*Albert Einstein*

4.1 Introduction

In Albert Einstein's quote above, he stresses the paramount importance of simplicity. In regression analysis, focusing on only two variables demonstrates the concepts simply. Thus, in Chap. 2, we calculated the least squares line by using two variables. We also plotted scatterplots and calculated correlation coefficients to further assess the linear relationship. From this analysis, we obtained a detailed understanding of the relationship between two variables. Upon understanding a linear relationship, other more complicated processes become easier to grasp.

Simple linear regression refers to linear regression with one predictor variable and one response variable. In the previous chapter, our primary focus entailed the calculation and fundamental programming concepts of linear regression. The goal of this chapter is to guide the reader through some of the understanding and analysis details of simple linear regression.

We begin with a discussion of the mathematical model of simple linear regression, the assumptions that are required to make inferences from the fitted model, and how to measure statistical significance of a fitted model. We then cover confidence intervals for the model coefficients, followed by an application. In the final discussion, we present a stock market case study and solve it in detail. The case study involves utilizing simple linear regression to calculate stock *betas*, which measure the performance of a stock as a function of the overall stock market. This case study makes use of R with descriptive graphs and the relevant source code.

© The Author(s), under exclusive license to Springer Nature Switzerland AG 2023
D. P. McGibney, *Applied Linear Regression for Business Analytics with R*,
International Series in Operations Research & Management Science 337,
https://doi.org/10.1007/978-3-031-21480-6_4

4.2 Simple Linear Regression Model

A mathematical model is an equation that describes the behavior of a system or concept mathematically. A mathematical model can be used to draw conclusions, such as the time it takes an object to fall to the ground or the speed of an orbiting satellite. Mathematical models are applicable to a wide array of subject areas. In regression analysis, the simple linear regression model provides one example of a simple mathematical model and is defined as

$$Y = \beta_0 + \beta_1 X + \varepsilon, \tag{4.1}$$

where β_0 and β_1 are parameters of the model. These parameters are the theoretical coefficients. The ε term is a random variable with a mean of 0 and is called the irreducible error or the error term. We include an error term in the equation to account for the deviation between the observed response value and the expected response value. Including the error term in the equation allows for measuring the error and dealing with it accordingly.

In using the least squares approach as outlined in Chap. 2, the simple linear regression model was automatically assumed although not explicitly stated.

4.3 Model Assumptions

Since we use the least squares criterion, the values $\hat{\beta}_0$ and $\hat{\beta}_1$ result in the best line that fits the data. For this model to be appropriate, four assumptions, as listed below, must hold true:

1. The relationship between the response and the predictor variable is linear.
2. Each value of ε_i is independent.
3. The random variable, ε, is normally distributed.
4. The variance of ε, denoted as σ^2, is the same for all observations.

Assumption 1, often referred to as the linearity assumption, is equivalent to assuming that the mean of each ε_i is 0. Mathematically, these assumptions can be summarized by the following:

$$\varepsilon_i \overset{\text{i.i.d.}}{\sim} N(0, \sigma^2).$$

This notation is read as: the ith error (ε_i) is independent and identically distributed (i.i.d.) and is distributed according to a normal (or Gaussian) distribution with a mean of 0 and a variance of sigma squared (σ^2). Chapter 7 provides a more in-depth discussion.

4.4 Model Variance

As mentioned in the model assumptions, the irreducible error of the model, ε, has a variance of σ^2 that implies the standard deviation is assumed to be σ. A large value of σ indicates the data are widely dispersed about the regression line, and a small value of σ indicates the data lie relatively close to the regression line. In Fig. 4.1, the variance in the error is depicted as relatively steady as x increases. Having a constant variance about the least squares line marks assumption 4 of the model as listed above. If this assumption were not met, some of the analysis associated with the variance may not be correct.

While the parameter σ^2 is the unknown theoretical value of the variance, an estimate can be obtained by dividing the SSE by the degrees of freedom:

$$s^2 = \frac{SSE}{n-2}. \tag{4.2}$$

The degrees of freedom are given by $n - 2$ when there are n observations in the data set and 2 coefficients in the model. The simple linear regression model has 2 coefficients (β_0 and β_1) as shown in Eq. (4.1). In simple linear regression, it is always the case that the degrees of freedom for the SSE are $n - 2$. From the approximation of the variance, an approximation of the standard deviation can be

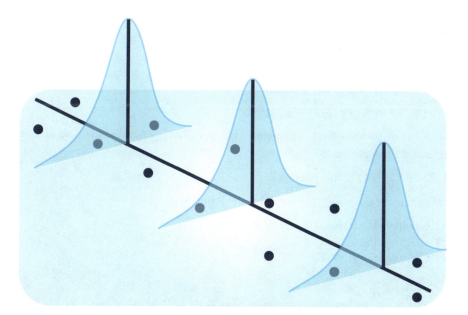

Fig. 4.1 Variance in the error

found by taking the square root:

$$s = \sqrt{s^2}. \tag{4.3}$$

Plugging in Eq. (4.2) for s^2, the value becomes

$$s = \sqrt{\frac{SSE}{n-2}}. \tag{4.4}$$

This standard deviation is often referred to as the residual standard error (RSE) or the root mean squared error ($RMSE$).

Recall from Eq. (3.18) in Chap. 3 that the SSE is

$$SSE = \sum_{i=1}^{n}(y_i - \hat{y}_i)^2. \tag{4.5}$$

If the simple regression equation is plugged in, the SSE becomes

$$SSE = \sum_{i=1}^{n}(y_i - \hat{\beta}_0 - \hat{\beta}_1 x_i)^2. \tag{4.6}$$

4.5   Application: Stock Revenues

Gabriele compiled financial information on the company Altryx. She then made predictions that Altryx had revenues of

$\hat{Y}$: 85 75 64 89 35 88 (in billions $)

for the years 2010 to 2015. Thereafter, she compared her revenue predictions with the actual revenue values:

Y: 83 80 73 90 47 80 (in billions $)

Investigate the estimates of Y by doing the following:

(a) Calculate the residuals for each year.
(b) Calculate the SSE.
(c) Calculate the $RMSE$.

Solution

(a) The calculated residuals are displayed in Table 4.1 within column 3. Recall from Chap. 4 the residuals are given by $y - \hat{y}$.

Table 4.1 Stock residuals

y	$\hat{y}$	$y - \hat{y}$	$(y - \hat{y})^2$
83	85	-2	4
80	75	5	25
73	64	9	81
90	89	1	1
47	35	12	144
80	88	-8	64

(b) The squared residuals are in column 4 of Table 4.1. From these squared residuals and Eq. (3.11), the SSE is

$$SSE = \sum_{i=1}^{n} (y_i - \hat{y}_i)^2$$

$$SSE = 4 + 25 + 81 + 1 + 144 + 64$$

$$SSE = 319.$$

(c) From Eq. (4.2), the variance estimate is

$$s^2 = \frac{SSE}{n - 2}$$

$$s^2 = \frac{319}{6 - 2} = 79.75.$$

From the variance estimate, the $RMSE$ is given by

$$s = \sqrt{s^2}$$

$$s = \sqrt{79.75} = 8.93.$$

4.6 Hypothesis Testing

Inferential statistics refers to the practice of drawing conclusions from data. The practice signifies an important topic not only in simple linear regression, but also in regression analysis in general. In fact, regression analysis addresses two major goals: prediction and inference. The purpose of prediction lies in estimating a variable. The value of prediction becomes somewhat obvious when different profit margins result from different business actions, but the value of inference may not

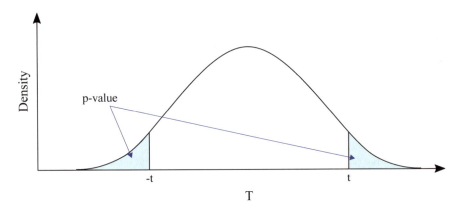

Fig. 4.2 A two-tailed *p*-value

be as obvious. One example of inference is interpreting how variables influence a predicted value, which can help explain a data set. Knowing that for every five sales calls a product is sold, for example, demonstrates the utility of inference.

Hypothesis testing exists as important tool in inferential statistics. In performing a hypothesis test, we assume a null hypothesis (H_0) and calculate a test statistic assuming the null hypothesis is true. If the statistic is within an acceptable range, then the null hypothesis is reasonable. However, if the test statistic is extreme, the null hypothesis is unlikely to be correct. Hypothesis testing of a mean is usually explained in introductory statistics, and the logic remains the same for more advanced hypothesis testing.

To efficiently conduct a hypothesis test, we introduce the concept of a *p*-value. Assuming H_0 is true, the probability that the test statistic will take on values as extreme as or more extreme than the observed test statistic is called a *p*-value.

Determining the results of a hypothesis test by using the *p*-value is called the *p*-value method, whereas comparing a test statistic with a critical value is referred to as the critical region method. In regression analysis, the *p*-value method is the preferred approach because *p*-values represent probabilities that offer a measure of the degree of rejection. Furthermore, test statistics are typically less intuitive than probabilities. A depiction of the *p*-value corresponding to a test statistic, *t*, is shown in Fig. 4.2. In this figure, the density area highlighted in blue represents the *p*-value, which results from a two-tailed hypothesis test using a *t*- distribution.

4.6.1   The qt Function

Traditionally, the critical value or *p*-value necessary to conduct a t-test was found using statistical tables that are widely used and easily obtained online. However, we may also easily and more accurately find this information using R. The qt function

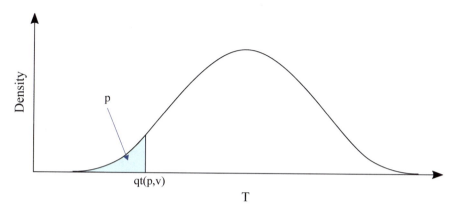

Fig. 4.3 The qt function

in R is used to find quantile values of the t-distribution and, therefore, allows us to find the t critical value.

The qt function has two required arguments and the following syntax:

```
qt(p, v)
```

The first argument (p) is a probability or a vector of probabilities, and the second argument (v) refers to the degrees of freedom. Figure 4.3 depicts the resulting t-value that can be found using the qt function. As with any function in base R, the full details on the qt function can be found in the R help files.

By entering a probability and the degrees of freedom, the qt function returns the t-value that corresponds to that probability. To find the critical value, $\alpha/2$ will be the first argument and $n - 2$ will be the second argument in the qt function. In which case the qt function will yield the negative critical value $(-t_{\alpha/2})$. Since the t-distribution is symmetrical, $1 - \alpha/2$ could be entered for p, which would result in the positive critical value $(t_{\alpha/2})$. By default, the probability for p will correspond to the area on the left-side of the quantile value.

4.6.2 The pt Function

Similarly, a pt function in R returns the probability corresponding to a quantile value. The syntax is

```
pt(t, v)
```

where t refers to the quantile (t-value), and v denotes the degrees of freedom. Figure 4.4 depicts the resulting cumulative probability density value (p) that can be found using the pt function.

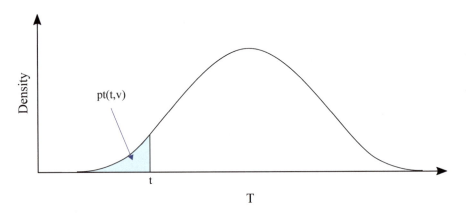

Fig. 4.4 The pt function

We can use the pt function to find the *p*-value corresponding to a test statistic. If we are looking for the *p*-value for a hypothesis test, the test statistic should be entered as t and the degrees of freedom should be entered as v. It is important to note the pt will return the area to the left of t; thus, the area will be less than 0.5 if t is negative but greater than 0.5 for a positive value of t. If there is a positive test statistic, the area more extreme than the t-value is

```
1 - pt(t, v)
```

When using *t*-tests in regression analysis, it will almost always be the case that the hypothesis test is two-tailed. For example, the area in the *t*-distribution that is more extreme than 2 is not only the tail above 2, but also the tail below −2. When finding the *p*-value in a two-tailed test, the results of the pt function will need to be multiplied by 2 when *t* is negative:

```
2*pt(t, v)
```

In the case when *t* is non-negative, we have

```
2*(1-pt(t, v))
```

Replacing pt(t, v) with 1-pt(t, v) accounts for the *t*-value being on the right-side, as the pt function returns the area to the left.

4.7   Application: Using the pt and qt Functions

Use your knowledge of the pt and qt functions to answer the following questions without the use of R. While the answers to these questions can be found using R, it is best to test your understanding of these concepts here.

Using the qt function: What will the following code return?

(a) `qt(0.5, 10)`
(b) `qt(0.5, 100)`
(c) `qt(0.025, 10000)`
(d) `qt(-0.025, 10000)`
(e) `qt(100, 10000)`

Using the pt function: What will the following code return?

(a) `pt(0, 5)`
(b) `pt(0, 25)`
(c) `pt(-1.96, 10000)`
(d) `pt(-100, 10000)`
(e) `pt(100, 10000)`

Solution
Using the qt function:

(a) Since having an area of 0.5 on the left of the graph represents the center of the *t*-distribution and the center of the distribution is at 0, this qt function will return 0. The degrees of freedom dictate the size of the distribution's tails that is irrelevant in this case since the distribution is symmetrical.
(b) The degrees of freedom dictate the size of the distribution's tails that is irrelevant in this case since the distribution is symmetrical. Having an area of 0.5 on the left-side of the graph corresponds to a *t*-value of 0.
(c) When the degrees of freedom are large, then the *t*-distribution emulates that of a standard normal distribution. Recalling the properties of the standard normal distribution from the empirical rule, having an area of 0.025 in the left-side of the graph corresponds to a *t*-value of -1.96.
(d) The first argument within the qt function represents a probability or an area, neither of which can be negative. Therefore, this function will return an error.
(e) Having a probability of 100 is impossible and will return an error.

Using the pt function:

(a) Since 0 represents the center of the distribution, the probability to the left of 0 is 50% or half. The degrees of freedom dictate the size of the distribution's tails that is irrelevant in this case since the distribution is symmetrical.
(b) The probability to the left of 0 is 50% or half. Even though the degrees of freedom change from the values in part (a), the distribution with 25 degrees of freedom is also symmetrical and centered about 0.
(c) When the degrees of freedom are large, then the *t*-distribution emulates that of a standard normal distribution. With this in mind, the area to the left of -1.96 in a standard normal distribution is 0.025 or 2.5%. For verification of this result, review the empirical rule from an introductory statistics source.
(d) As in (c), we note that the large degrees of freedom result in the distribution being approximately equal to a standard normal distribution. The area to the

left of -100 in a standard normal distribution is approximately 0 since -100 is an extreme value in the left tail.

(e) As in (c) and (d), the large degrees of freedom result in the distribution being approximately equal to a standard normal distribution. The area to the left of 100 in a standard normal distribution is approximately 1 since 100 is an extreme value in the right tail.

4.8 Hypothesis Testing: Student's t-Test

When testing the significance of a slope parameter β_1, it may prove valuable to check if the parameter is zero, since a zero β_1 means that there is no relationship between the response and the predictor variables. Therefore, a test can be created to check if β_1 is zero or nonzero. Setting up the hypotheses, the null hypothesis would stipulate that the parameter is 0, and the alternative hypothesis would therefore be that the parameter would be nonzero. Using statistical symbols, this can be written as:

$$H_0 : \beta_1 = 0$$

$$H_1 : \beta_1 \neq 0,$$

where H_0 and H_1 denote the null and alternative hypotheses, respectively.

To find a test statistic for this test, $\hat{\beta}_1$ is used as the estimate of β_1. The value of $\hat{\beta}_1$ is divided by the standard error of $\hat{\beta}_1$:

$$t_1 = \frac{\hat{\beta}_1}{s_{\hat{\beta}_1}}. \tag{4.7}$$

The standard error of $\hat{\beta}_1$ is given by

$$s_{\hat{\beta}_1} = \frac{s}{\sqrt{SS_{xx}}}. \tag{4.8}$$

Note that the standard deviation of the error is used to determine the standard error of $\hat{\beta}_1$.

The value of t_1 represents a student's t-distribution with $n-2$ degrees of freedom. Therefore, the null hypothesis can be rejected if the p-value corresponding to t_1 lies below the level of significance (α). Although somewhat arbitrary, it is common to set α to 0.05. This value can usually be assumed if there is no mention of α.

4.9 Employee Churn Application: Testing for Significance with t

Karen from human resources at Global IT Enterprises is trying to find the number of employees that will leave the company by department. The number of employees in a department (X) can be used to predict the number of employees that will churn (Y). The data are shown here in Table 4.2.

Using the formulas from Chap. 2, we calculate the regression equation as

$$\hat{y} = 2.1818 + 1.6452x.$$

Investigate the relationship between X and Y by doing the following:

(a) Calculate the SSE.
(b) Calculate the standard deviation of the error.
(c) Calculate the standard error of the slope.
(d) Calculate the test statistic for the slope.
(e) Compute the p-value.
(f) Make the decision.
(g) Interpret the slope.

Solution

(a) The predicted values are easily obtained by plugging in each value of X into the regression equation:

$$\hat{y}_1 = 2.1818 + 1.6452(2) = 5.4722$$

$$\hat{y}_2 = 2.1818 + 1.6452(2) = 5.4722$$

$$\hat{y}_3 = 2.1818 + 1.6452(5) = 10.4078$$

$$\hat{y}_4 = 2.1818 + 1.6452(1) = 3.827$$

$$\hat{y}_5 = 2.1818 + 1.6452(3) = 7.1174$$

$$\hat{y}_6 = 2.1818 + 1.6452(.5) = 3.0044$$

$$\hat{y}_7 = 2.1818 + 1.6452(7) = 13.6982.$$

Table 4.2 Churn data

Variable							
Employees (in 100s) (X)	2	2	5	1	3	0.5	7
Churns (Y)	7	5	12	4	8	1	12

Plugging in the predicted values ($\hat{y}$) and the actual values of y into Eq. (3.11), the SSE is

$$SSE = \sum_{i=1}^{n}(y_i - \hat{y}_i)^2.$$

$$SSE = (7 - 5.4722)^2 + (5 - 5.4722)^2 + (12 - 10.4078)^2 + (4 - 3.827)^2$$
$$+(8 - 7.1174)^2 + (1 - 3.0044)^2 + (12 - 13.6982)^2$$
$$= 12.80266.$$

(b) From Eq. (4.2), the variance of the error is

$$s^2 = \frac{SSE}{n-2}$$

$$s^2 = \frac{12.80266}{7-2}$$

$$s^2 = 2.560532.$$

The standard deviation is therefore

$$s = \sqrt{2.560532}$$

$$s = 1.600166.$$

(c) To calculate the standard error of the slope, the standard deviation of the error and the sum of squares of X (SS_{xx}) are required. The sum of squares of X is

$$SS_{xx} = \sum_{i=1}^{n} x_i^2 - \left(\sum_{i=1}^{n} x_i\right)^2 / n$$

$$SS_{xx} = 92.25 - (20.5)^2/7$$

$$SS_{xx} = 32.21429,$$

and the standard deviation of the error was calculated in the previous part. Using Eq. (4.8), the standard deviation of the slope estimate is therefore

$$s_{\hat{\beta}_1} = \frac{s}{\sqrt{SS_{xx}}}$$

$$s_{\hat{\beta}_1} = \frac{1.600166}{\sqrt{32.21429}}$$

$$s_{\hat{\beta}_1} = 0.2819297.$$

(d) The test statistic for the slope, as stated in Eq. (4.7), is calculated to be

$$t_1 = \frac{\hat{\beta}_1}{s_{\hat{\beta}_1}}$$

$$t_1 = \frac{1.6452}{0.2819297}$$

$$t_1 = 5.835497.$$

(e) The p-value corresponding to $t_1 = 5.835497$ and 5 degrees of freedom can be found using the R function:

```
2*pt(-5.835497, 5)
```

```
## [1] 0.002089956
```

(f) Since the p-value of 0.002089956 is less than $\alpha = 0.05$, the null hypothesis should be rejected at the $\alpha = 0.05$ level of significance. The slope is therefore statistically significant.

(g) Since the slope is 1.6452: for every 100 employees in a department, it is expected approximately 1.6452 will churn.

4.10   Coefficient Confidence Interval

Confidence intervals give us an interval estimate for our approximation assuming a confidence level of $1-\alpha$. The confidence interval for β_1 in a simple linear regression model is:

$$\hat{\beta}_1 \pm t_{\alpha/2}s_{\hat{\beta}_1}, \tag{4.9}$$

where $\hat{\beta}_1$ is the point estimate of β_1, $t_{\alpha/2}$ is the t-value such that $\alpha/2$ is in the upper tail with degrees of freedom of $n - 2$, and $s_{\hat{\beta}_1}$ is the standard error of the slope.

The confidence interval of a coefficient can be calculated using the confint function in R. If there is a regression object, reg, available in the R environment, the confidence interval can be calculated as

```
confint(reg)
```

This R function requires a regression object as input and defaults to a 95% confidence interval. The optional level argument can be specified to return a different level of confidence.

In the previous sections, a test statistic was used to conclude the hypothesis test for t. Additionally, confidence intervals of the coefficients can be used to test the hypotheses used in the previous t-test. For instance, when $\alpha = 0.05$ if the value of $\beta_1 = 0$ is outside the 95% confidence interval, then H_0 is rejected since the null hypothesized value of β_1 is not included in the confidence interval. In the application below, we conduct an equivalent hypothesis test using a confidence interval for β_1.

4.11   Employee Churn Application: Confidence Interval Hypothesis Testing

Again using Karen's churn data (Table 4.2), find out if the number of employees in each department (X) is a significant variable in determining the number of employees that will churn (Y).

Recall that the regression equation is

$$\hat{y} = 2.1818 + 1.6452x.$$

Investigate the relationship between X and Y by doing the following:

(a) Set up a statistical test for β_1.
(b) Calculate the $t_{\alpha/2}$ critical value.
(c) Find a 95% confidence interval for β_1.
(d) Conclude the test using the confidence interval.

Solution

(a) The null hypothesis states that β_1 is 0 and the alternative hypothesis is nonzero:

$$H_0 : \beta_1 = 0$$

$$H_1 : \beta_1 \neq 0.$$

(b) Since α is split into two equal parts at the beginning and end of the distribution, the probability value of 0.025 is the probability argument for the first critical value. Keeping in mind that the qt function takes as arguments areas to the right, we plug in 0.975 as the probability argument to get the right-side critical value. The values corresponding to $\pm t_{\alpha/2}$ with 5 degrees of freedom and $\alpha = 0.05$ can be found using the R code:

```
qt(0.025, 5)
qt(0.975, 5)
```

```
## [1] -2.570582
```

```
## [1] 2.570582
```

From this result, the positive value is returned:

$$t_{\alpha/2} = 2.570582.$$

(c) Using Eq. (4.9) and the values calculated in the previous application, the confidence interval becomes

$$\hat{\beta}_1 \pm t_{\alpha/2} s_{\hat{\beta}_1}$$

$$1.6452 \pm 2.570582 \times 0.2819297$$

or written as

$$0.9204766 \le \beta_1 \le 2.369923.$$

As confirmation, the 95% confidence intervals for both β_0 and β_1 can be calculated using the confint function in R once a regression object is created using the lm function:

```
confint(reg)
```

```
##                   2.5 %    97.5 %
## (Intercept) -0.4490947 4.812731
## x            0.9205094 2.369956
```

Note that the confidence interval limits for β_1 calculated by R are consistent with the earlier calculation when ignoring minor rounding errors.

(d) Since 0 is not included in the confidence interval for β_1, the null hypothesis is rejected. The estimate $\hat{\beta}_1$ is therefore said to be statistically significant.

4.12 Hypothesis Testing: F-test

In addition to the t-test, the F-test, which signifies the ratio of two chi-square distributions, represents another important statistical test in regression analysis. In general, the F-test determines if any significance exists among predictor variables. When there is only one predictor variable, as in simple linear regression, the setup and results are the same as for the t-test. For models with more than one predictor variable, once the F-test deems the model significant, the t-test may be used to determine whether individual predictor variables are significant.

The hypotheses for an F-test in simple linear regression are as follows:

$$H_0 : \beta_1 = 0$$

$$H_1 : \beta_1 \neq 0.$$

The test statistic for F is given as

$$F = \frac{MSR}{MSE},$$
(4.10)

where MSR is the mean squares due to regression that is

$$MSR = \frac{SSR}{p}.$$
(4.11)

In this equation, p represents the number of predictor variables, which is one for the simple linear regression case. The number of predictors denotes the degrees of freedom for the regression.

The rejection rule holds that one should reject H_0 if F is at or more extreme than the critical value for a given α. An equivalent method constitutes verifying that the p-value of the test statistic occurs at or below α. In particular, reject H_0 if p-value $\leq \alpha$ or if $F \geq F_\alpha$ where the critical value F_α is based on an F distribution with p degrees of freedom in the numerator and $n - p - 1$ degrees of freedom in the denominator.

The null hypothesis of the F-test posits no linear relation between the predictor variables and the response variable. The alternative hypothesis posits a linear relation between at least one predictor variable and the response variable.

4.13 The qf Function

We use the qf function in R to find quantile values of the F distribution. Similar to the qt function, the qf function can return the critical value of an F distribution if the significance level is specified. In contrast to the t-distribution, the F distribution

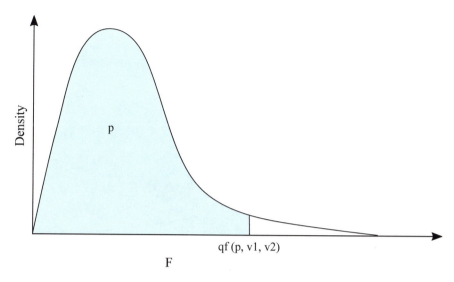

Fig. 4.5 The qf function

requires two degrees of freedom values since this distribution is the ratio of two χ^2 distributions.

The qf function has three required arguments. The syntax is as follows:

```
qf(p, v1, v2)
```

The first argument (p) refers to a probability or a vector of probabilities, whereas the second and third arguments (v1 and v2) are the degrees of freedom in the numerator and denominator, respectively. By entering a probability and the degrees of freedom, the qf function returns the F-value that corresponds to that probability, as depicted in Fig. 4.5. To find the critical value, $1 - \alpha$ will be the first argument. In simple linear regression, the degrees of freedom in the numerator are 1, which is attributed to the number of predictor variables, and the degrees of freedom in the denominator are $n - 2$. Note that the F distribution shape varies with different degrees of freedom combinations.

4.14 The pf Function

The pf function in R returns the probability corresponding to a quantile value for the F distribution. The syntax is

```
pf(f, v1, v2)
```

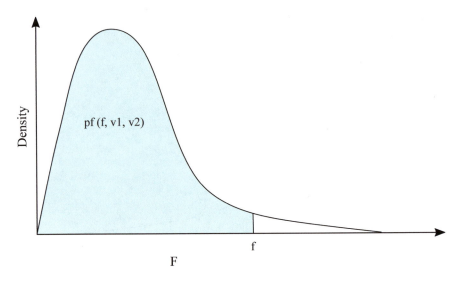

Fig. 4.6 The pf function

where f denotes the quantile (F-value) and the second and third arguments (v1 and v2) are the degrees of freedom in the numerator and denominator, respectively. Note that pf will return the area to the left of f by default. Figure 4.6 shows a depiction of the pf function.

4.15   Employee Churn Application: Testing for Significance with F

Again using Karen's churn data, find out if the number of employees in each department (X) is a significant variable in determining the number of employees that will churn (Y).

Recall that the regression equation is

$$\hat{y} = 2.1818 + 1.6452x.$$

Investigate the relationship between X and Y by doing the following:

(a) Calculate the F critical value with $\alpha = 0.05$.
(b) Calculate the F-test statistic.
(c) Make the F-test decision.
(d) Find the p-value from the F-test statistic.
(e) Comment on the p-value.
(f) Interpret the results.

Solution

(a) The F critical value is 1 degree of freedom in the numerator (v_n) and $n - 2$ degrees of freedom in the denominator (v_d). The qf R function gives

```
qf(0.95, 1, 5)
```

```
## [1] 6.607891
```

(b) To find the F-statistic, one must first find the SSR. Using Eq. (3.17), the SSR is

$$SSR = \sum_{i=1}^{n} (\hat{y} - \bar{y})^2$$

$$SSR = (5.4722 - 7)^2 + (5.4722 - 7)^2 + (10.4078 - 7)^2 + (3.827 - 7)^2$$
$$+ (7.1174 - 7)^2 + (3.0044 - 7)^2 + (13.6982 - 7)^2$$
$$= 87.19386$$

From this value of SSR, the MSR is

$$MSR = SSR/v_n$$

$$MSR = 87.19386/1$$

$$MSR = 87.19386.$$

Using the value of the MSE from the previous application, the F- statistic is

$$F = MSR/MSE$$

$$F = 87.19386/2.560532$$

$$F = 34.05302.$$

(c) Since the F-statistic is more extreme than the critical value, the null hypothesis is rejected. More specifically,

$$F > F_\alpha$$

$$34.05302 > 6.607891.$$

(d) The p-value from F with 1 degree of freedom in the numerator (v_n) and $n - 2$ degrees of freedom in the denominator (v_d) can be found with the pf function:

```
1 - pf(34.05302, 1, 5)
```

```
## [1] 0.002089957
```

(e) There are two important findings from the p-value:

1. The p-value corresponding to the F-statistic is equivalent to the one for the t statistic. In simple linear regression, the F-test is equivalent to the t-test that results in the same p-values.
2. The hypothesis test can be concluded by comparing the p-value with α. This test gives the same result as if the critical value were compared to the F-test statistic (as in part c).

(f) The F-value is more extreme than the critical value, and therefore, the model is significant.

4.16 Cautions About Statistical Significance

In practice, if a t-test or F-test shows that rejecting the null hypothesis is the appropriate course of action, it does not enable us to definitively conclude that a cause-and-effect relationship exists between x and y. Concluding a test as shown in this chapter may suggest correlation, but correlation between variables does not imply causation. Also, finding statistical significance in the aforementioned tests does not enable us to conclude that there is a linear relationship between x and y. A nonlinear model may be a better fit regardless of the significance.

4.17 Case Study: Stock Betas

4.17.1 Problem Statement

In studying the stock market, data analysis can provide insights about the selection of stocks, sectors, and the overall market. Noting the individual stock performance relative to the overall market is of particular importance. One such measure of this performance is referred to as *beta*. As shown in the next section, beta can be derived by fitting a regression model. This case demonstrates how regression analysis can be used in finance, particularly stock market analysis.

The beta value for a particular stock can be found by fitting a simple linear regression model, which can be written as

$$R_a = \alpha + \beta R_b + \varepsilon, \tag{4.12}$$

where:

- R_a is the return of an asset (or stock).
- R_b is the return of a market benchmark.
- ϵ is the error of the model.

The remaining model parameters α and β denote the intercept and slope of the linear regression model. The estimate of β can be interpreted as "correlated relative volatility," but in simpler terms, this means that for every percentage increase in the market benchmark (R_b), the R_a increases by β percent.

For the market benchmark in this case study, we will use the S&P 500, which is commonly used as a measure of the market's total performance. Using several individual stocks, we will fit a regression model and calculate the resulting coefficients.

As an analyst working for a hedge fund, you are tasked with analyzing data on seven stocks to obtain the beta values for each. The seven stocks, which are of particular interest to management, are listed here, along with their tickers:

- Apple Incorporate (AAPL)
- Caterpillar Incorporated (CAT)
- Johnson & Johnson (JNJ)
- McDonald's Corporation (MCD)
- Procter & Gamble Company (PG)
- Microsoft Corporation (MSFT)
- Exxon Mobil Corporation (XOM)

Specifically, you will need to use data from the 36 months prior to September 1, 2021. Using the data, you will then find each stock's beta value and create some visualizations. From the calculated beta values, you will then need to make some investment recommendations to management.

4.17.2 Descriptive Statistics

The case study data consist of 36 monthly returns of a selection of stocks. The first step in analyzing a clean data set is to fully understand the data set of interest. After loading in the data, examine the first 6 observations using the head function.

```
df = read.csv("Betas.csv")
head(df)
```

```
##           X            AAPL           CAT           JNJ           MCD
## 1 2018-09-01 -0.004824971  0.09823557  0.03267315  0.037627656
## 2 2018-10-01 -0.030477500 -0.20440695  0.01317225  0.057445226
## 3 2018-11-01 -0.184044589  0.12540913  0.04936063  0.065630391
## 4 2018-12-01 -0.113616340 -0.06338909 -0.11591696 -0.052224994
## 5 2019-01-01  0.055154105  0.04792657  0.03122828  0.006814199
## 6 2019-02-01  0.040314724  0.03802575  0.02675073  0.028302710
##           PG          MSFT           XOM           SPY
## 1  0.003375568  0.02207889  0.071434774  0.001412313
## 2  0.065481313 -0.06610134 -0.062808643 -0.064890545
## 3  0.075159052  0.03819886 -0.002258965  0.018549474
## 4 -0.027404354 -0.08009031 -0.133569056 -0.093343088
## 5  0.049499294  0.02815791  0.074644220  0.086372988
## 6  0.029641118  0.07277600  0.078466127  0.032415740
```

The data set from "Betas.csv" was assigned to the df dataframe using the read.csv function. Note the first column corresponds to the date of each return, while the next 7 columns consist of the monthly returns for 7 stocks, and the last column represents the S&P 500 (denoted as SPY) index. The first date value is September 1, 2018, and the second date value is October 1, 2018, indicating each date corresponds to the first of the month. On closer inspection, the dates range from September 1, 2018 to August 1, 2021. The dataframe can be modified using simple indexing to exclude the first column. Here we indicate that column 1 should be excluded by placing a negative one within square brackets "[ ]" that indexes all columns except the first one.

```
df = df[-1]
```

After loading in a data set, the head function gives the analyst a glimpse of the first 6 observations of the data. Data sets can also be understood by finding descriptive statistics and graphs that summarize the information. The summary command provides a quick and powerful way to compute descriptive statistics in R.

```
summary(df)
```

```
##      AAPL                CAT                 JNJ
## Min.   :-0.18404   Min.   :-0.20441   Min.   :-0.11592
## 1st Qu.:-0.03588   1st Qu.:-0.03730   1st Qu.:-0.01916
## Median : 0.05580   Median : 0.03986   Median : 0.02056
## Mean   : 0.03300   Mean   : 0.01726   Mean   : 0.01066
## 3rd Qu.: 0.10116   3rd Qu.: 0.06148   3rd Qu.: 0.04187
## Max.   : 0.21438   Max.   : 0.18699   Max.   : 0.14421
##      MCD                 PG                  MSFT
## Min.   :-0.143098   Min.   :-0.08600   Min.   :-0.080090
## 1st Qu.:-0.009394   1st Qu.:-0.01084   1st Qu.: 0.005981
```

```
## Median : 0.017790   Median : 0.01786   Median : 0.034708
## Mean   : 0.014158   Mean   : 0.01829   Mean   : 0.030299
## 3rd Qu.: 0.051348   3rd Qu.: 0.05453   3rd Qu.: 0.062072
## Max.   : 0.134321   Max.   : 0.09660   Max.   : 0.136326
##      XOM                 SPY
## Min.   :-0.2512614   Min.   :-0.1299871
## 1st Qu.:-0.0599777   1st Qu.:-0.0004336
## Median : 0.0001581   Median : 0.0220387
## Mean   :-0.0006238   Mean   : 0.0153075
## 3rd Qu.: 0.0722371   3rd Qu.: 0.0428491
## Max.   : 0.2238609   Max.   : 0.1336104
```

From the summary, the data become easier to understand, but one common statistic that remains missing is the standard deviation, which represents an important measure of spread that will help assess the variability of the stock returns. To get the standard deviation of the Microsoft returns, the sd function can be used.

```
sd(df$MSFT)
```

```
## [1] 0.05361294
```

Rather than finding the standard deviation of each stock return separately, the standard deviations can be calculated using a single line of code. Specifically, the sapply function can apply the sd function across every variable in our dataframe.

```
sdevs = sapply(df, sd)
sdevs
```

```
##       AAPL        CAT        JNJ        MCD         PG
## 0.09515141 0.08226207 0.05387847 0.05494649 0.04376807
##       MSFT        XOM        SPY
## 0.05361294 0.10281086 0.05413432
```

Even though the mean is included in the summary, the mean can be calculated in a similar manner.

```
means = sapply(df, mean)
means
```

```
##          AAPL           CAT           JNJ           MCD
## 0.0329988211 0.0172605185 0.0106559689 0.0141578462
##            PG          MSFT           XOM           SPY
## 0.0182918257 0.0302987475 -0.0006237706 0.0153074727
```

4.17.3 Plots and Graphs

As a simple yet informative graph, a bar graph can represent the mean monthly returns for each stock. The `barplot` function in R can generate this visualization with relative ease.

```
barplot(means, xlab = "Stock Ticker",
        ylab = "Mean Monthly Returns",
        main = "Mean Monthly Stock Returns")
```

From this barplot, we see that Apple (AAPL) has the highest mean return among the stocks, closely followed by Microsoft (MSFT) and then Proctor & Gamble (PG). Depending on the purpose of our analysis of these stocks, we may find importance in noting the lowest two stocks are Exxon Mobil (XOM) and Johnson & Johnson (JNJ), respectively.

Previously, a vector of standard deviations was calculated and printed. A barplot of the standard deviations provides a useful visualization.

```
barplot(sdevs, xlab = "Stock Ticker",
        main = "Std. Dev. of Monthly Returns")
```

Recall the standard deviation is a measure of variability and is often associated with risk. As shown in the barplots above, Exxon Mobil has the largest standard deviation (Fig. 4.8) and the lowest mean of the stocks listed (Fig. 4.7). While Exxon Mobil's

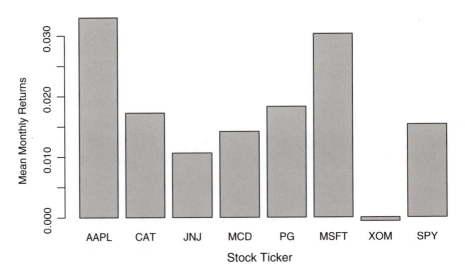

Fig. 4.7 Barplot of means

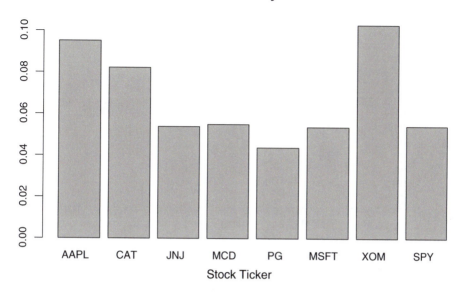

Fig. 4.8 Barplot of standard deviations

low mean indicates it did not perform well, the variability shown by the standard deviation would indicate it is a risky investment.

In contrast to Exxon Mobil's performance, it is often the case that a large standard deviation is consistent with a large mean, which is shown from Apple's relatively large mean and standard deviation. This relation between high mean performance and high standard deviation reinforces the notion that with greater risk comes greater reward.

The S&P 500 often has low variability since it is an index consisting of 500 stocks, as indicated by a low standard deviation. However, from Fig. 4.8, we observe that Proctor & Gamble has an even lower standard deviation. To investigate this unlikely result, we use another plot to observe the variability further.

As discussed in Chap. 2, the boxplot function is an important R tool to analyze data. Boxplots are great for visualizing the variability and the center of each variable simultaneously.

```
boxplot(df, xlab = "Stock Ticker", ylab = "Monthly Returns",
        main = "Boxplot of Monthly Returns of Stocks")
```

From Fig. 4.9, several conclusions can be made. For example, we can see that Exxon Mobil has the highest variability, which is in agreement with the standard deviation. From the standard deviations, we noted that Proctor & Gamble had the lowest standard deviation. However, the S&P 500 appears to have a lower IQR variability indicating that the middle 50% of the observations have lower variability than the

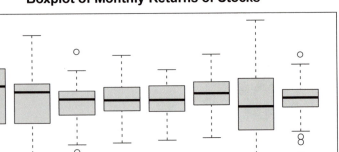

Fig. 4.9 Boxplot of returns

middle 50% of observations from Proctor & Gamble. The S&P 500 generally has lower variability since it represents an index of 500 stocks, but it appears a few outliers were able to drive the standard deviation higher than Proctor & Gamble.

4.17.4   Finding Beta Values

The beta value of Microsoft can be found easily by using the lm function. The syntax for the lm function is as follows:

```
lm(y ~ x, data = df)
```

where y is the response, x is the predictor variable, and df is the dataframe where the y and x variables are located.

To find the beta value for Microsoft, use the following:

```
regression = lm(MSFT ~ SPY, data = df)
summary(regression)
```

```
##
## Call:
## lm(formula = MSFT ~ SPY, data = df)
##
## Residuals:
```

```
##         Min        1Q    Median        3Q       Max
## -0.058379 -0.031883  0.003456  0.022605  0.084847
##
## Coefficients:
##              Estimate Std. Error t value Pr(>|t|)
## (Intercept) 0.019050   0.006323   3.013  0.00486 **
## SPY         0.734825   0.113871   6.453 2.24e-07 ***
## ---
## Signif. codes:
## 0 '***' 0.001 '**' 0.01 '*' 0.05 '.' 0.1 ' ' 1
##
## Residual standard error: 0.03647 on 34 degrees of freedom
## Multiple R-squared:  0.5505, Adjusted R-squared:  0.5373
## F-statistic: 41.64 on 1 and 34 DF,  p-value: 2.235e-07
```

From the summary, the value of the slope coefficient (beta) for Microsoft is 0.734825. The regression equation is therefore

$$\text{MSFT} = 0.019050 + 0.734825 \times \text{SPY}, \qquad (4.13)$$

where MSFT is the predicted value of the Microsoft returns, as predicted by the S&P 500 returns (SPY). This regression equation makes use of the model in Eq. (4.12).

Since the slope coefficient represents the desired result, we may attain it by typing:

```
regression$coefficients[2]
```

```
##       SPY
## 0.7348248
```

The r-squared value can also be found from the regression by typing:

```
summary(regression)$r.squared
```

```
## [1] 0.5505208
```

4.17.5   Finding All the Betas

We may find all of the beta values by getting the slope from each model. This process can be automated with a loop in order to find the beta of multiple stocks. Notice that the dataframe (df) has 8 variables with names given by

```
nm = names(df)
nm
```

```
## [1] "AAPL" "CAT"  "JNJ"  "MCD"  "PG"   "MSFT" "XOM"  "SPY"
```

To contain the betas, an object with the first 7 names from df each having a value of 0 is created with the setNames() function.

```
betas = setNames(numeric(7), nm[1:7])
```

Similarly, an object to contain the r_squares can be created with the code:

```
r_squares = setNames(numeric(7), nm[1:7])
```

Noting that the paste() function can be used to create a formula by concatenating text strings. More specifically, the formula for making the first stock (Apple) the response and the S&P 500 (SPY) the predictor variable would be

```
paste(nm[1], "~ SPY")
```

```
## [1] "AAPL ~ SPY"
```

A for loop can be used to repeat an operation for different indices (such as stocks). For example, if the indices are 1 through 7, a simple for loop can go through each value and print out each index.

```
for (i in 1:7){
  print(i)
}
```

```
## [1] 1
## [1] 2
## [1] 3
## [1] 4
## [1] 5
## [1] 6
## [1] 7
```

Using a simple for loop, one can find each value of beta and R^2.

```
for (i in 1:7){
  regression = lm(paste(nm[i], "~ SPY"), data = df)
  betas[nm[i]] = regression$coefficients[2]
  r_squares[nm[i]] = summary(regression)$r.squared
}
```

A barplot of the betas provides a visual depiction:

```
barplot(betas, xlab = "Stock Ticker", ylab = "Beta",
        main = "Beta Values by Stock")
```

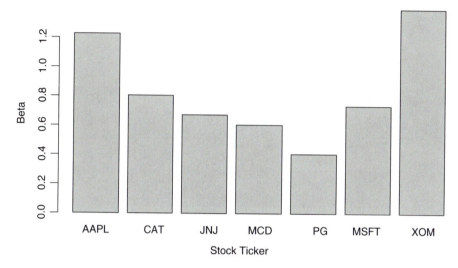

Fig. 4.10 Barplot of betas

If a stock beta is 1, then the stock increases at the same rate as the market benchmark. If a stock beta is above 1, then the stock is considered to be more volatile than the market. From the stock beta plot, Fig. 4.10, we notice Apple and Exxon Mobil are more volatile than the market.

The names of the stocks with betas above 1 can be confirmed using the following code.

```
names(betas[betas > 1])
```

```
## [1] "AAPL" "XOM"
```

Stocks with betas below 1 are considered to be less volatile than the market. The names of the stocks with betas below 1 can be found using the following code.

```
names(betas[betas < 1])
```

```
## [1] "CAT"  "JNJ"  "MCD"  "PG"  "MSFT"
```

A visual representation of the R^2 values is shown below.

```
barplot(r_squares, xlab = "Stock Ticker",
        ylab = "Percent Return",
        main = "Percent of Return Explained by Market")
```

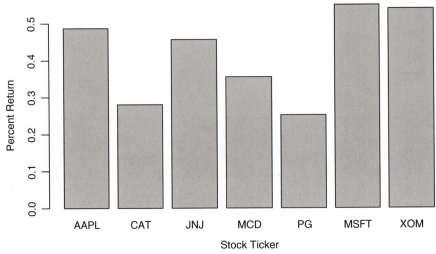

Fig. 4.11 Barplot of R^2 values

By looking at the stock R^2 values in Fig. 4.11, we note that Exxon Mobil is the stock most correlated with the market. To investigate this matter further, we plot a scatterplot to see the relationship.

```
plot(df$SPY, df$XOM, xlab="S&P 500 Benchmark Return",
    ylab="Exxon Mobil Return")
```

In Fig. 4.12, we note that there is a clear relationship with the S&P 500. However, the vertical axis has a larger range than the horizontal axis. This indicates further evidence that Exxon Mobil, the vertical axis in the figure, is more variable than the market.

4.17.6   Recommendations and Findings

From this analysis, we draw several conclusions. In taking the mean of each variable, Apple showed the best performance followed by Microsoft. However, when looking at the variability, Apple had the second highest variability, which demonstrates the risk–reward relationship. Looking at the other extreme, Exxon Mobil had the lowest average return in the data set and the highest standard deviation, making it a poor investment. Proctor & Gamble had the third largest mean and a standard deviation that indicated a low risk. Having a low standard deviation and a high mean makes Proctor & Gamble a sound investment.

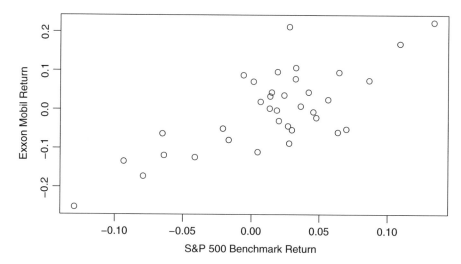

Fig. 4.12 Scatterplot of returns

From this analysis, we can use the following insights to formulate some investment strategies for management:

- It is probably a safe prediction that Proctor & Gamble will not be affected as much as other stocks if the S&P 500 does not perform well. Therefore, investing in Proctor & Gamble is a sound choice given that the return is relatively high.
- If Apple continues to have high variability and high returns, then it may be advantageous to buy Apple when the price is on the downswing, since the variability of Apple's stock is high and the long-term mean return is high.
- Since Microsoft and Exxon Mobil are correlated with the market, if we are able to accurately predict the market's performance, it would be possible to accurately predict the returns of Microsoft and Exxon Mobil.

Despite the analysis done here, keep in mind that financial analysts warn that the past does not guarantee future performance.

4.17.7   Case Conclusion

Analytics and the broader field of AI are revolutionizing global markets. AI relies on some basic programming commands such as `for` loops in order to repeat the same script for a number of different indices. Also, creating functions is important to implement a programmer's logic. Both `for` loops and functions were discussed in this case.

Using free data is often the first step for students learning about stock market analysis. The data for this case were downloaded for free from Yahoo Finance using the quantmod package. For additional details on downloading data using the quantmod package, see the Appendix. Many base R commands were used to model and analyze the data, such as: barplot, boxplot, c, head, lm, mean, names, paste, plot, print, sapply, sd, subset, summary, and setNames.

By analyzing stock market indicators, analysts can better understand the current market trends. This analysis conveyed some basic stock market analysis to demonstrate the use of these tools and techniques. The lessons learned here can be expanded to generate more plots and get even better insights on more complex data.

Problems

1. **Social Media Influencers F-Statistic**
   Using the social media influencer data from Table 3.2, calculate the F-statistic by answering the following questions. Assume we are trying to predict the number of clicks based on the amount of pay.
   Without the use of a computer:

   a. Find $\hat{y}_i$ for each x_i.
   b. Find the residual corresponding to each y.
   c. Calculate the SST.
   d. Calculate the SSR.
   e. Calculate the SSE.
   f. Calculate the F-value for the regression.

   While using a computer (with R):

   g. Verify your results from parts a through f.

2. **Social Media Influencers t-Statistic**
   Using the data from Table 3.2, answer the following questions.
   Without the use of a computer:

   a. Calculate the mean square error (MSE) and the root mean square error ($RMSE$).
   b. Calculate the standard error of the slope estimate ($\hat{\beta}_1$).
   c. Calculate the t-value for the slope estimate ($\hat{\beta}_1$).

   While using a computer (with R):

   d. Verify your results from parts a through c.

3. **Social Media Influencers Hypothesis Tests**
   Using the data from Table 3.2, answer the following questions:

   a. Specify the null and alternative hypotheses for the t-test.
   b. Find the p-value for the t-test using R.
   c. Compare the p-value to the level of significance ($\alpha = 0.05$) and make the decision.
   d. Specify the null and alternative hypotheses for the F-test.
   e. Find the p-value for the F-test using R.
   f. How do the p-values of the F and t-tests differ?
   g. Interpret the results of both tests.

4. **Accounting Analytics F-Statistic**
   Using the data from the previous chapter case study, calculate the F-statistic by answering the questions. Assume you are predicting the itemized deductions using the adjusted gross income. The data are given in Table 3.1.
   Without the use of a computer:

   a. Find $\hat{y}_i$ for each x_i.
   b. Find the residual corresponding to each y.
   c. Calculate the SST.
   d. Calculate the SSR.
   e. Calculate the SSE.
   f. Calculate the F-value for the regression.

   While using a computer (with R):

   g. Verify your results from parts a through f.

5. **Accounting Analytics t-Statistic**
   Using the data from Table 3.1, answer the following questions.
   Without the use of a computer:

   a. Calculate the mean square error (MSE) and the root mean square error ($RMSE$).
   b. Calculate the standard error of the slope estimate ($\hat{\beta}_1$).
   c. Calculate the t-value for the slope estimate ($\hat{\beta}_1$).

   While using a computer (with R):

   d. Verify your results from parts a through c.

6. **Accounting Analytics Critical Region Method for t**
   Using the data from Table 3.1, answer the following questions:

   a. Specify the null and alternative hypotheses for the t-test.
   b. Find the critical values for the t-test using R with $\alpha = 0.05$.
   c. Compare the critical value to the test statistic for t and make the decision.
   d. Interpret the results of the t-test.

7. **Sales Calls F-Statistic**
   Using Abdullah's calls and sales data from Table 2.4, answer the following
   questions:

   a. Find $\hat{y}_i$ for each x_i.
   b. Find the residual corresponding to each y_i.
   c. Calculate the SST.
   d. Calculate the SSR.
   e. Calculate the SSE.
   f. Calculate the F-value for the regression.

8. **Sales Calls t-Statistic**
   Using the data from Table 2.4, answer the following questions.

   a. Calculate the mean square error (MSE) and the root mean square error
      ($RMSE$).
   b. Calculate the standard error of the slope estimate ($\hat{\beta}_1$).
   c. Calculate the t-value for the slope estimate ($\hat{\beta}_1$).

9. **Automotive Stocks F-Statistic**
   Using the automotive stock data from Table 3.3 to predict price from earnings,
   answer the following questions:

   a. Find $\hat{y}_i$ for each x_i.
   b. Find the residual corresponding to each y.
   c. Calculate the SST.
   d. Calculate the SSR.
   e. Calculate the SSE.
   f. Calculate the F-value for the regression.

10. **Automotive Stocks Test Statistic for t**
    Using the data from Table 3.3, answer the following questions:

    a. Calculate the mean square error (MSE) and the root mean square error
       ($RMSE$).
    b. Calculate the standard error of the slope estimate ($\hat{\beta}_1$).
    c. Calculate the t-value for the slope estimate ($\hat{\beta}_1$).

11. **Automotive Stocks F-Tests**
    Using the data from Table 3.3, answer the following questions:

    a. Find the critical value for the F-test using R. Specify the function, argu-
       ments, and output.
    b. Compare the critical value to the test statistic for F and make the decision.
    c. Find the p-value for the F-test using R. Specify the function, arguments, and
       output.
    d. Compare the p-value to the level of significance ($\alpha = 0.05$) and make the
       decision.

Chapter 5
Multiple Regression

Prediction is very difficult, especially about the future.
—Niels Bohr

5.1 Introduction

In this chapter, we build upon the coverage of regression analysis by considering situations involving two or more predictor variables. For instance, while the weight of a person may be predicted using their height, we could use both the height and age of that person to predict their weight. Using more than one predictor variable to predict a response is called multiple regression analysis, which enables us to consider more predictor variables and thus obtain better estimates than those possible with simple linear regression.

Since the calculations of the multiple regression coefficients are identical to those of the simple linear regression, it is imperative to first fully understand the simple linear regression case prior to learning multiple linear regression. As seen in Chap. 3, calculating coefficients and model statistics aids in the understanding of simple linear regression. In multiple regression, most of the calculations build on the knowledge gained from working with simple linear regression, but the calculations become more tedious due to the number of predictor variables involved. Therefore, we limit the calculations done by hand or calculator when working with multiple regression and instead rely upon R and its functions for such calculations.

We begin by specifying the multiple regression model and equation. Then, we discuss the limitations of using R^2 in multiple regression (referred to as multiple R^2). Next, we introduce a measure of model fit that is more appropriate for multiple regression (adjusted R^2). We use a website marketing application to illustrate these concepts while keeping the number of predictor variables limited to two for simplicity. In the final discussion of the chapter, we analyze a housing data set in a case study, providing detailed analyses for the reader.

© The Author(s), under exclusive license to Springer Nature Switzerland AG 2023
D. P. McGibney, *Applied Linear Regression for Business Analytics with R*,
International Series in Operations Research & Management Science 337,
https://doi.org/10.1007/978-3-031-21480-6_5

5.2   Multiple Regression Model

In the previous chapters, we discussed in detail the simple linear regression model and the utility of having a mathematical model. In multiple regression, the aim is to capture the effect of multiple predictor variables in predicting a response variable. The equation that describes how the response variable Y is related to p predictor variables $X_1, X_2, ..., X_p$ and an error term is

$$Y = \beta_0 + \beta_1 X_1 + \beta_2 X_2 + \cdots + \beta_p X_p + \varepsilon, \qquad (5.1)$$

where $\beta_0, \beta_1, \beta_2, \ldots, \beta_p$ are the parameters and ε is a random variable representing the error in the model. If only β_0 and β_1 are nonzero and the remaining coefficients: $\beta_2, \beta_3, \ldots, \beta_p$ are zero, then the resulting model would be the simple linear regression model:

$$Y = \beta_0 + \beta_1 X_1 + \varepsilon \qquad (5.2)$$

as mentioned in the previous chapter. As is the case for both the simple linear regression model and the multiple regression model, the assumptions for the linear regression model must be adhered to. A detailed discussion of the assumptions for linear regression and how to deal with violations of the linear model is given in Chap. 8.

5.3   Multiple Regression Equation

The mathematical model given from Eq. (5.1) is called a multiple linear regression model. When we use and fit the multiple linear regression model with a real data set, then the corresponding multiple linear regression equation is

$$\hat{Y} = \hat{\beta}_0 + \hat{\beta}_1 X_1 + \hat{\beta}_2 X_2 + \cdots + \hat{\beta}_p X_p. \qquad (5.3)$$

The sample statistics or coefficients $\hat{\beta}_0, \hat{\beta}_1, \hat{\beta}_2, \ldots, \hat{\beta}_p$ can be calculated using a given data set and the set of matrix equations. These sample statistics are point estimators of the corresponding parameters: $\beta_0, \beta_1, \beta_2, \ldots, \beta_p$.

5.4   Website Marketing Application: Modeling

A recent startup company is developing a website for news and entertainment. They notice that their monthly revenue (in thousands of dollars) is dependent on the number of ads and the amount of money (in thousands of dollars) spent on additional marketing. The data are shown in Table 5.1.

Table 5.1 Website
marketing data

Ads	Marketing	Revenue
4	14	118
5	8	113
6	13	149
4	14	129
5	10	131
5	5	126
5	11	129
5	9	125
6	12	150
5	11	124
5	11	126
6	11	159
5	12	130
4	9	114
6	8	136
3	9	80
6	7	137
5	8	128
6	9	149
5	10	136

Investigate the relationship between the response and the predictors by doing the
following:

(a) Specify the multiple linear regression model that results from predicting revenue
    by the number of ads and the amount spent on marketing.
(b) Fit the model using R.
(c) Explicitly state the regression equation.

Solution

(a) Since it is useful to predict the revenue based on the number of ads and
    the amount spent on marketing, it follows that the revenue generated is an
    appropriate response variable. The multiple linear regression model can be
    written as

$$Y = \beta_0 + \beta_1 X_1 + \beta_2 X_2 + \varepsilon,$$

where:

- Y is the amount of revenue generated.
- X_1 is the number of ads.
- X_2 is the amount spent on additional marketing.
- ε is the error.

The model, therefore, has parameters β_0, β_1, and β_2. Fitting the data to a model generates approximations of the parameters that are referred to as $\hat{\beta}_0$, $\hat{\beta}_1$, and $\hat{\beta}_2$.

(b) The regression analysis for this problem can quickly be carried out using R. As a first step, we use the c function to enter in the data.

```
Revenue = c(118, 113, 149, 129, 131, 126, 129, 125, 150, 124,
            126, 159, 130, 114, 136, 80,  137, 128, 149, 136)
Ads= c(4, 5, 6, 4, 5, 5, 5, 5, 6, 5, 5, 6, 5, 4, 6, 3, 6, 5, 6,
       5)
Marketing = c(14, 8, 13, 14, 10, 5, 11, 9, 12, 11, 11, 11, 12,
              9, 8, 9, 7, 8, 9, 10)
```

Once the data set is read into R, a regression object can be created using the lm function. In the code below, the object is named reg. To view a summary of the regression model created, use the summary function as shown:

```
reg = lm(Revenue ~ Ads + Marketing)
summary(reg)
```

```
##
## Call:
## lm(formula = Revenue ~ Ads + Marketing)
##
## Residuals:
##     Min      1Q  Median      3Q     Max
## -10.849  -4.726  -1.352   4.816  10.096
##
## Coefficients:
##              Estimate Std. Error t value Pr(>|t|)
## (Intercept)   12.5506    12.5314   1.002  0.33061
## Ads           18.6143     1.8930   9.833 1.98e-08 ***
## Marketing      2.2783     0.6714   3.393  0.00346 **
## ---
## Signif. codes:
## 0 '***' 0.001 '**' 0.01 '*' 0.05 '.' 0.1 ' ' 1
##
## Residual standard error: 6.747 on 17 degrees of freedom
## Multiple R-squared:  0.8558, Adjusted R-squared:  0.8389
## F-statistic: 50.45 on 2 and 17 DF,  p-value: 7.091e-08
```

The output of the summary function is quite extensive and can be used to reconstruct the regression equation by looking at the coefficients.

(c) Upon inspection of the summary output, the resulting multiple linear regression equation is

$$\hat{Y} = 12.55 + 18.61X_1 + 2.28X_2,$$

which is often written in terms of the actual variable names, making it easier to interpret:

$$\text{Revenue} = 12.55 + 18.61(\text{Ads}) + 2.28(\text{Marketing}).$$

5.5 Significance Testing: t

We conduct a separate t-test for each of the predictor variables in the model. Each of these t-tests constitutes a test for individual significance.

The hypotheses for the t-test of the ith variable are as follows:

$$H_0 : \beta_i = 0$$

$$H_1 : \beta_i \neq 0,$$

and the corresponding test statistic can be calculated by dividing the coefficient estimate ($\hat{\beta}_i$) by the corresponding standard error of the coefficient ($s_{\hat{\beta}_i}$):

$$t_i = \frac{\hat{\beta}_i}{s_{\hat{\beta}_i}}. \qquad (5.4)$$

In general, a t-test statistic can be calculated for every coefficient in the multiple regression model. The t-value corresponds to a p-value that can be calculated using the pt function in R as shown in the previous chapter. For multiple regression models, we use a t-distribution with $n - p - 1$ degrees of freedom. We reject the null hypothesis and deem the coefficient estimate ($\hat{\beta}_i$) is deemed statistically significant when the p-value is less than α. As discussed in Chap. 4, a two-tailed test is most relevant for a coefficient test since the p-value takes into consideration extreme values on both sides of the T-distribution. Alternatively, the test can be concluded by comparing t_i with the critical values ($t_i \leq -t_{\alpha/2}$ or $t_i \geq t_{\alpha/2}$).

5.6 Coefficient Interpretation

In multiple regression analysis, the ith regression coefficient $\hat{\beta}_i$ can be interpreted as the impact of one unit increase in x_i on the response variable when all other predictor variables are held constant.

While small correlations among x_i variables often occur, large correlations may hinder the interpretation of the regression coefficients. If little or no correlation among predictor variables is assumed, then the interpretation of the coefficients becomes quite convenient.

5.7   Website Marketing Application: Individual Significance Tests

From the regression summary, a coefficient summary is constructed and shown below in Table 5.2. This table displays several statistics that indicate important details about the fitted regression coefficients. Most importantly, we see the estimated values of $\hat{\beta}_0$, $\hat{\beta}_1$, and $\hat{\beta}_2$, as given in the second column. The corresponding standard errors of $\hat{\beta}_0$, $\hat{\beta}_1$, and $\hat{\beta}_2$ are given in the third column, with the resulting t-values and p-values in the fourth and fifth columns, respectively.

Using the estimate and the standard error of the estimate from Table 5.2, do the following:

(a) Verify the calculation of the t-value using the estimate and the standard error.
(b) Verify the p-value for each coefficient using the t-value and the pt function.
(c) Conclude the t-test for each coefficient.
(d) Interpret the coefficients.

Solution

(a) For the website marketing application, the t-values can be calculated using Eq. (5.4):

$$t_0 = \frac{\hat{\beta}_0}{s_{\hat{\beta}_0}} = \frac{12.551}{12.531} = 1.002$$

$$t_1 = \frac{\hat{\beta}_1}{s_{\hat{\beta}_1}} = \frac{18.614}{1.893} = 9.833$$

$$t_2 = \frac{\hat{\beta}_2}{s_{\hat{\beta}_2}} = \frac{2.278}{0.6714} = 3.393.$$

Table 5.2 Website marketing regression

	Estimate	Std. error	t-value	p-value
(Intercept)	12.550623	12.5313599	1.001537	0.3306108
Ads	18.614323	1.8930117	9.833179	0.0000000
Marketing	2.278313	0.6713902	3.393426	0.0034566

(b) Since the *t*-distribution is symmetric, extreme *t*-values can exist on both the right- and left-side of the distribution. Thus, we use a two-tailed test, which considers both extreme values. In this case, each *p*-value is two-tailed and can be calculated using the formula given here. Recall that the negative of the absolute *t*-value is taken in order to return the extreme area on the left.

```
2*pt(-t, n-p-1)
```

From the code above and noting that the degrees of freedom are $n - p - 1 = 20 - 2 - 1 = 17$, the *p*-values corresponding to t_0, t_1, and t_2 are given by the R code.

```
2*pt(-1.002, 17)
2*pt(-9.833, 17)
2*pt(-3.393, 17)
```

The code above yields near the values listed in the summary output, although slight discrepancies occur due to rounding. Note that, in this case, each *p*-value is two-tailed with $n - p - 1 = 20 - 2 - 1 = 17$ degrees of freedom since there are 20 observations and 2 predictors.

(c) Since the *p*-value of the intercept is larger than $\alpha = 0.05$, the null hypothesis cannot be rejected and is, therefore, not statistically significant. The slope estimates of $\hat{\beta}_1$ and $\hat{\beta}_2$ are both statistically significant since both of the corresponding *p*-values are less than $\alpha = 0.05$.

(d) Since the intercept is not statistically significant, it should not be interpreted. The coefficient corresponding to the "Ads" variable is $\hat{\beta}_1 = 18.61$. Therefore, the coefficient can be interpreted as:

For each additional ad, an increase of $18,610 in revenue occurs when money on additional marketing is held constant.

In a similar way, $\hat{\beta}_2 = 2.28$ can be interpreted as:

The revenue increases by $2,280 for every $1000 spent via other marketing means assuming number of ads is held constant.

5.8 Significance Testing: *F*

In simple linear regression, the *F*- and *t*-tests provide the same conclusion since the null and alternative hypotheses are the same. In fact, the resulting *p*-values of the tests are also equivalent despite using different statistics. In multiple regression, however, these tests have different conclusions. We use the *F*-test to determine if the model is significant. Once the *F*-test deems the model significant, we should use the *t*-test to determine whether each individual predictor variable is significant.

The hypotheses for an F-test are as follows:

$$H_0 : \beta_1 = \beta_2 = \ldots = \beta_p = 0$$

$$H_1 : \text{At least one coefficient is nonzero.}$$

The null hypothesis of the F-test states that there is no linear relation between the predictor variables and the response variable. The alternative hypothesis states that there is a linear relation between at least one predictor variable and the response variable.

The F-test takes into account the entire regression model when determining significance. The test statistic is given by the ratio of the mean squares due to regression (MSR) and the mean squared error (MSE):

$$F = \frac{MSR}{MSE} \tag{5.5}$$

or in terms of multiple R^2:

$$F = \frac{R^2}{1 - R^2} \left(\frac{n - p - 1}{p} \right). \tag{5.6}$$

The rejection rule stipulates that one should reject H_0 if F is at or more extreme than the critical value for a given level of α. An equivalent method entails verifying if the p-value of the test statistic is at or below α. In particular, reject H_0 if p-value $\leq \alpha$ or if $F \geq F_\alpha$, where the critical value F_α is based on an F-distribution with p degrees of freedom in the numerator and $n - p - 1$ degrees of freedom in the denominator.

5.9   Multiple R^2 and Adjusted R^2

For multiple regression, many of the components are similar to simple linear regression but take into consideration multiple predictor variables. Particularly, while we discussed the coefficient of determination (R^2) between two variables, we would like to find the coefficient of determination for the multiple predictor case (multiple R^2). Multiple R^2 is given by

$$R^2 = SSR/SST. \tag{5.7}$$

In the website marketing application, both variables were significant in predicting the revenue. In practice, however, a new predictor variable will not always aid in the prediction of the response variable. In some of the simple linear regression examples, the R^2 value was very low, and we easily showed that the predictor variable did not aptly predict the response. In multiple regression, including an additional predictor variable, even new variables that are not statistically significant, will always result in a higher multiple R^2. This concept can be explained intuitively by considering that predictor variables with a low multiple R^2 still contribute to reducing the error in fitting the model.

Mathematically, the prediction errors become smaller, thus reducing the sum of squares due to error, SSE. Because

$$SSR = SST - SSE, \tag{5.8}$$

when SSE becomes smaller, SSR becomes larger, causing

$$R^2 = \frac{SSR}{SST} \tag{5.9}$$

to increase. Even though increasing the SSR causes the R^2 to increase, the addition of predictor variables does not always increase the true explained variation. Therefore, it is better to avoid using the multiple R^2 as a means of model evaluation for multiple regression.

Also consider the case in which over a hundred predictor variables exist, but each predictor variable consists of random noise with no predictive power. If a response variable were introduced with less than a hundred observations, then a model could be fit to predict the response variable with the random noise predictors. In fact, since the number of observations would be limited, the random noise could be easily fit to predict the response. The error or SSE would be at or near zero, implying a great fit, but the model fit would be misleading.

To counteract this limitation of R^2, we introduce a similar statistic without the pitfalls mentioned previously. The adjusted multiple coefficient of determination is known as the adjusted R^2 and written as R_a^2, which compensates for the number of predictor variables in the model and the number of observations. This adjustment allows for a more useful measure since it is affected by the number of predictor variables. The R_a^2 amplifies the unexplained variation ratio $(1 - R^2)$ by the ratio of total degrees of freedom $(n-1)$ with the degrees of freedom due to error $(n-p-1)$. The result is the equation

$$R_a^2 = 1 - (1 - R^2)\frac{n-1}{n-p-1}, \tag{5.10}$$

which raises the unexplained variation by taking p, the number of predictor variables, into account.

5.10   Website Marketing Application: Multiple R^2 and Adjusted R^2

In the website marketing application, the SSR is 4593.13 and the SST is 5366.95. Recalling that there are 20 observations and two predictors, do the following:

(a) Calculate the multiple R^2.
(b) Calculate the adjusted R^2.
(c) Interpret the multiple R^2 and the adjusted R^2.

Solution

(a) The multiple R^2 is given by Eq. (5.9):

$$R^2 = \frac{SSR}{SST} = \frac{4593.13}{5366.95} = 0.8558.$$

(b) From Eq. (5.10),

$$R_a^2 = 1 - (1 - R^2)\frac{n-1}{n-p-1}$$

$$= 1 - (1 - 0.8558)\frac{20-1}{20-2-1} = 0.8389,$$

which agrees with the summary output from the modeling application above.

(c) The multiple R^2 indicates that 85.58% of the variation in Y is explained by the two predictor variables. The adjusted R^2 indicates that 83.89% of the variation in Y is explained by the two predictor variables taking into account the number of observations and the number of predictors.

5.11   Correlations in Multiple Regression

A correlation matrix shows the linear correlation between each pair of variables under consideration in a multiple regression model. While correlations between the predictor and response variables serve to predict the response variable, correlations among predictor variables may be problematic for modeling the regression. For instance, if predictor variables X_1 and X_2 have a correlation coefficient of 0.99, then there would be little benefit to using both X_1 and X_2 to predict the response variable, Y. In fact, using both X_1 and X_2 could be a detriment to the predictive power of the model since noise is likely the reason for the difference between the predictors. We expect some amount of random noise to be present in all random variables whether they be a response or a predictor. Therefore, if X_1 contains most of the same information about Y that X_2 contains, using a multiple regression model with both predictor variables would compound the effects of the noise.

The effect of compounding noise as mentioned above is referred to as multicollinearity. Multicollinearity exists between two predictor variables if they have a high linear correlation. If two predictor variables in the regression model are highly correlated with each other, it is important to investigate the regression output.

Since inspecting a correlation matrix will indicate if one or more high correlations exist between predictor variables, using additional, more advanced techniques, such as the inspection of variance inflation factors (VIFs), may not be necessary to find multicollinearity.

5.12   Case Study: Real Estate

5.12.1   Problem Statement

Real estate represents a popular investment not only among wealthy investors, but also among lower and middle-class investors. Because of this widespread interest, analysis of real estate data can be particularly interesting.

The board of directors for an investment firm has sought your expertise in analyzing their data. You are tasked with analyzing a data set with 90 similar cities in a large state. You will need to fit a regression model, analyze the model, interpret the coefficients, and test a claim made by a local politician. The directors of the investment firm need your input so that they can determine which city they should invest in. First, you will need to understand the data set using some basic R commands.

5.12.2   Data Description

The data consist of 150 similar cities in a large state. The state of interest remains unknown as the investment firm directors would like to retain some secrecy. The firm leaders requested that the model predicts price as a function of the remaining variables in the data set.

The variables for each city are:

- Price—average price of home
- Rooms—average number of rooms
- Income—average income
- Tax rate—property tax rate
- Commercial—percent of commercial property

As a first step, we load the data file into R and observe the first 6 observations.

```
df = read.csv("HousePrices.csv")
head(df)
```

```
##   price.in.usd rooms income Tax.Rate X.Commercial
## 1       107496 4.494  61168    1.937        21.26
## 2       102012 4.235  56311    3.975        20.22
## 3       118100 4.415  64746    1.557        13.45
## 4       102439 4.526  56074    2.771        20.52
## 5        97648 4.043  47908    4.275        12.93
## 6       104024 4.261  51624    2.155        22.82
```

Notice that R replaces several of the characters in the variable names with "."
and the "%" character with "X." The variable names can easily be renamed using
the names function:

```
names(df) = c('Price', 'Rooms', 'Income', 'TaxRate', 'Commercial')
```

The plot function in base R can be used to create scatterplots of two variables;
however, it can also be used to quickly create a matrix of scatterplots. If the first
argument of the plot function is a dataframe with more than 3 variables, then a
matrix of scatterplots is created (Fig. 5.1).

```
plot(df)
```

The scatterplots prove particularly interesting with regard to the response variable. Upon a cursory glance, the average price is positively correlated with "Rooms"
and "Income" and negatively correlated with "Tax Rate" and "Commercial." Using
the cor function in base R, the correlations can be verified.

```
cor(df)
```

```
##                    Price       Rooms        Income
## Price          1.0000000  0.50143372  0.4048224196
## Rooms          0.5014337  1.00000000  0.4022251672
## Income         0.4048224  0.40222517  1.0000000000
## TaxRate       -0.3778801  0.02102138 -0.0780820791
## Commercial    -0.1364745 -0.21719890 -0.0005592479
##                  TaxRate     Commercial
## Price        -0.377880100 -0.1364744745
## Rooms         0.021021381 -0.2171989031
## Income       -0.078082079 -0.0005592479
## TaxRate       1.000000000 -0.0023067737
## Commercial   -0.002306774  1.0000000000
```

While this information verifies our conclusions from the scatterplots, the correlation matrix can be difficult to read and understand. Using the corrplot package, the

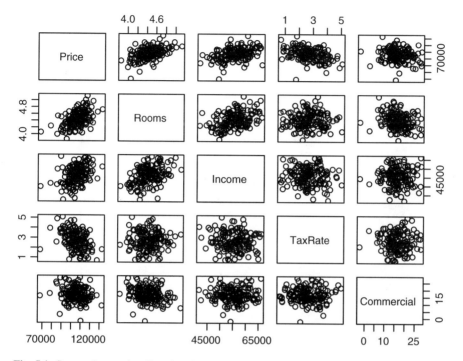

Fig. 5.1 Scatterplot matrix of housing data

matrix may be color-coded and much more easily read. If you have not yet installed the `corrplot` package, you may easily install it using the `install.packages` command.

```
install.packages("corrplot")
```

To generate a correlation matrix, one can simply use the `cor` command and then run `corrplot` on the resulting correlation matrix. Several additional options are available to make the graph more visually appealing.

```
library(corrplot)
c = cor(df)
corrplot(c)
```

Using the `corrplot` command, an elegant matrix of correlations can be created. The corrplot command has the following optional arguments:

- add—Set this argument to TRUE to superimpose the corrplot on another corrplot.
- method—Set this argument to "number" to display the numeric values of the correlation coefficients.
- type—Set this to "upper" or "lower" to get an upper or lower diagonal correlation matrix.

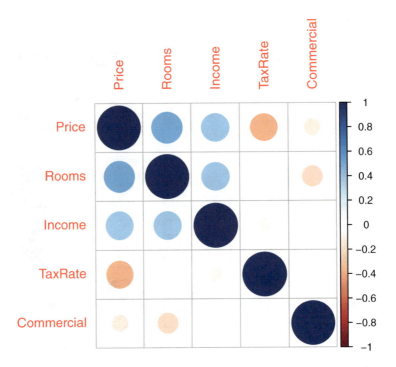

Fig. 5.2 Correlation plot of housing data 1

- diag—Toggles whether to include the diagonal elements of the matrix.
- tl.pos—Set this to "n" to hide the axis labels.
- cl.pos—Set this to "n" to hide the color legend.

```
corrplot(c)
corrplot(c, add = TRUE, method = "number", type = "lower",
             diag = FALSE, tl.pos = "n", cl.pos = "n")
```

Figure 5.2 is generated by using the `corrplot` command with the correlation matrix c as the only input, whereas Fig. 5.3 is generated by using a second `corrplot` function and specifying the options mentioned in the above bulleted list as shown in the code above.

Note that some of the correlations between the response and other variables are relatively weak (particularly "Price" with "Commercial"). A t-test can better accomplish testing for individual significance between predictor and response variables.

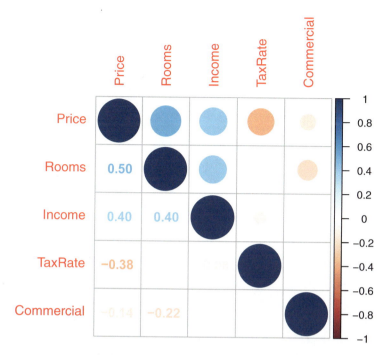

Fig. 5.3 Correlation plot of housing data 2

5.12.3   Simple Linear Regression Models

If one were only familiar with simple linear regression, a simple linear regression model could be fit for each predictor variable. In particular, a model could be generated and fit for each $i = 1, 2, 3, 4$, where

$$Y = \beta_0 + \beta_i X_i + \varepsilon.$$

Here, Y is the average price (Price), X_1 is the average number of rooms (Rooms), X_2 is the average income (Income), X_3 is the tax rate (Tax Rate), X_4 is the percent of commercial real estate (Commercial), ε is the error, and the unknown parameter values are $\beta_0, \beta_1, \beta_2, \beta_3, \beta_4$. The models can be fit from the dataframe using the `lm` command as shown in the previous chapters.

```
reg1 = lm(Price ~ Rooms, data = df)
reg2 = lm(Price ~ Income, data = df)
reg3 = lm(Price ~ TaxRate, data = df)
reg4 = lm(Price ~ Commercial, data = df)
```

The slope values are calculated from the previous code chunk and can be accessed from the fitted regression objects (reg1, reg2, reg3, reg4) as done in the stock beta case study:

```
reg1$coefficients[2]
reg2$coefficients[2]
reg3$coefficients[2]
reg4$coefficients[2]
```

```
##     Rooms
## 20468.44

##     Income
## 0.7573866

##    TaxRate
## -3878.652

## Commercial
##   -248.5123
```

Note that the second coefficient, referenced using the square brackets, represents the slope coefficient for each regression model.

The corresponding p-values of the slopes are available in the coefficients of the summary for each regression:

```
s = summary(reg1)
s$coefficients
```

```
##              Estimate Std. Error  t value     Pr(>|t|)
## (Intercept) 14358.41  12922.066 1.111155 2.683037e-01
## Rooms       20468.44   2903.053 7.050661 6.317987e-11
```

Since the p-value for the slope is the fourth column and second row shown in the output above, the p-values from all four fitted models can be returned by specifying "[2, 4]" from the summary coefficients as follows:

```
summary(reg1)$coefficients[2, 4]
summary(reg2)$coefficients[2, 4]
summary(reg3)$coefficients[2, 4]
summary(reg4)$coefficients[2, 4]
```

```
## [1] 6.317987e-11

## [1] 2.772951e-07

## [1] 1.866897e-06

## [1] 0.0958564
```

From the output above, all of the slope values are significant in each simple linear regression model. Furthermore, the R^2 values are

```
summary(reg1)$r.squared
summary(reg2)$r.squared
summary(reg3)$r.squared
summary(reg4)$r.squared
```

```
## [1] 0.2514358

## [1] 0.1638812

## [1] 0.1427934

## [1] 0.01862528
```

and the R_a^2 values are

```
summary(reg1)$adj.r.squared
summary(reg2)$adj.r.squared
summary(reg3)$adj.r.squared
summary(reg4)$adj.r.squared
```

```
## [1] 0.2463779

## [1] 0.1582317

## [1] 0.1370014

## [1] 0.01199437
```

which are slightly less than the R^2 values, since R_a^2 take into account the low number of observations.

5.12.4   Multiple Regression Model

As discussed previously, the directors of the investment firm would like us to create a model to predict the average price based upon all of the other variables in the data. Therefore, our model will be a multiple regression model with four predictor variables:

$$Y = \beta_0 + \beta_1 X_1 + \beta_2 X_2 + \beta_3 X_3 + \beta_4 X_4 + \varepsilon,$$

where Y, X_1, X_2, X_3, X_4 are the same values from the simple linear regression models. However, the error (ε) and the unknown parameter values β_0, β_1, β_2, β_3, β_4 may potentially be different values.

The model can be fit from the dataframe using the `lm` command:

```
reg = lm(Price ~ Rooms + Income + TaxRate + Commercial,
         data = df)
summary(reg)
```

```
##
## Call:
## lm(formula = Price ~ Rooms + Income + TaxRate
##                + Commercial, data = df)
##
## Residuals:
##     Min      1Q Median      3Q     Max
## -32241   -3881     191    4288   22422
##
## Coefficients:
##                  Estimate Std. Error t value Pr(>|t|)
## (Intercept) 20083.5355 12355.8105    1.625  0.10624
## Rooms        16937.6420  2859.3548    5.924 2.19e-08 ***
## Income           0.3910     0.1283    3.048  0.00274 **
## TaxRate      -3801.8121   641.4735   -5.927 2.16e-08 ***
## Commercial     -85.7450   116.5882   -0.735  0.46325
## ---
## Signif. codes:
## 0 '***' 0.001 '**' 0.01 '*' 0.05 '.' 0.1 ' ' 1
##
## Residual standard error: 7132 on 145 degrees of freedom
## Multiple R-squared:  0.4391, Adjusted R-squared:  0.4236
## F-statistic: 28.37 on 4 and 145 DF,  p-value: < 2.2e-16
```

In the summary, the values of the coefficients prove significantly different from those of the simple linear regression models. In fact, the coefficient for "Commercial" is not significant, even though the coefficient was significant in the simple linear regression case. Comparing the R^2 values from simple linear regression, even the highest value of 0.2514358 is not higher than the R^2 value from the multiple regression model, indicating a better model fit using multiple variables. Earlier in this chapter, the reader was warned about comparing models using R^2 alone. Therefore, we compare R^2_a and note that, when taking into consideration the number of observations and the number of variables, the variation in Y is better explained using a combination of variables than using a single variable.

The `reg` object was created with a formula that specified four variables. For convenience, the `reg` object could have been equivalently defined using a coding shorthand as shown below.

```
reg = lm(Price ~ ., data = df)
```

Note the "." on the right-side of the formula. Recall that the right-side of the formula is where we denoted our predictor variables previously. Having "." on the right-side of the formula tells R to include all of the variables except the response variable in the dataframe as predictor variables in the model.

The regression equation can therefore be expressed as

$$\hat{Y} = 20083.5355 + 16937.6420X_1 + 0.3910X_2 - 3801.8121X_3 - 85.7450X_4,$$

where the variables are defined above. Additionally, the regression equation can be expressed as

$$\text{Price} = 20083.5355 + 16937.6420 \times \text{Rooms} + 0.3910 \times \text{Income}$$
$$-3801.8121 \times \text{TaxRate} - 85.7450 \times \text{Commercial}.$$

Model Interpretation

The summary function provides a myriad of information we need for our analysis. First, note that the F-value corresponds to a small p-value, which indicates that the model has significance and, therefore, should be investigated further. The value of multiple R^2 is 0.4391, but since this model is a multiple regression, the number of variables used should be taken into account. Thus, the R_a^2 value of 0.4236 is preferred in the interpretation of this regression. Recall that the R^2 value represents the percent of variation explained by our model, in this case about 43.91% of the variation.

Coefficient Interpretation

Each coefficient estimate is tested when fitting a regression model in R. The coefficients can be interpreted as follows:

- Intercept—the intercept coefficient p-value is 0.106, which is larger than 0.05 (p-value $> \alpha$). The conclusion of the hypothesis test results in an insignificant intercept. There is not enough evidence to support that the intercept is nonzero for this model.
- Rooms—the coefficient for rooms has a small p-value, much smaller than 0.05 (p-value $< \alpha$). The conclusion holds that the coefficient is not 0. The coefficient can be interpreted as: "For every room, an increase in price of $16,937.64 is expected." This interpretation assumes that all of the other variables remain unchanged or constant.
- Income—the coefficient for income has a small p-value, also smaller than 0.05 (p-value $< \alpha$). The conclusion holds that the coefficient is not 0. The coefficient may be interpreted as: "For every dollar of income, an increase in price of $0.391

is expected." Like the "Rooms" coefficient, this interpretation assumes that all of the other variables remain unchanged or constant.

- Tax Rate—the coefficient for tax rate has a small p-value, much smaller than 0.05 (p-value $< \alpha$). The conclusion holds that the coefficient is not 0. We can further interpret the coefficient as: "For every percentage increase in the tax rate, a decrease in price of $3801.81 is expected." This interpretation assumes that all of the other variables remain unchanged or constant.

- Commercial—the coefficient for price has a p-value larger than 0.05 (p-value $> \alpha$). Therefore, the commercial variable is not nonzero for this model, and the coefficient for commercial should not be interpreted.

Confidence Interval

As another task given by the directors of the investment firm: we must test the claim of a local politician. The politician states that, for every percent decrease in the real estate tax, the property value increases by $5000. Can this claim be refuted?

From the tax rate coefficient of the fitted model, the estimate for the increase in property value for a 1% decrease is $3801.81. While the estimate in R should be trusted over a random claim, a simple hypothesis can be conducted. We may set up the null and alternative hypotheses as follows:

$$H_0 : \beta_3 = -\$5000$$

$$H_1 : \beta_3 \neq -\$5000.$$

Note that the null hypothesis states that the coefficient is negative, which implies home values decrease in value by $5000 for every 1% increase in tax.

This test is easy to conclude using the confidence interval for β_3. The confint function in R will calculate 95% confidence intervals for each coefficient.

```
confint(reg)
```

```
##                    2.5 %        97.5 %
## (Intercept) -4337.2243748 44504.2954268
## Rooms        11286.2428127 22589.0411426
## Income           0.1374199     0.6445348
## TaxRate      -5069.6585228 -2533.9656892
## Commercial    -316.1768576   144.6868108
```

Since −$5000 falls within the 95% confidence interval, the claim by the politician cannot be rejected at the 5% level of significance.

Was the politician correct in his statement? Our data indicate that the best estimate is $3801.81, while the politician claims the value as $5000. A significant difference between the estimate and the claim seems clear. In fact, if we change the level of significance to $\alpha = 0.1$, we attain a different result. We run the `confint` function again while specifying the level of significance at $\alpha = 0.1$, which is equivalent to specifying a confidence level of 0.9 as shown in the R code:

```
confint(reg, level = 0.9)
```

```
##                         5 %              95 %
## (Intercept)    -370.6470269 40537.7180789
## Rooms          12204.1795192 21671.1044361
## Income             0.1786043     0.6033503
## TaxRate        -4863.7267296 -2739.8974824
## Commercial      -278.7486327   107.2585859
```

The above represents a scenario where an estimate was made possibly with the intent to deceive by positing a claim that the value is close to one end of the confidence interval. However, as demonstrated above, the claim is difficult to disprove.

5.12.5   Case Conclusion

In this analysis, we analyzed a housing data set by creating and observing a scatterplot matrix and a correlation plot of relevant variables. We then fit simple linear regression models. While these models proved statistically significant, the amount of variation explained by each model was relatively small. We then specified and fit a multiple regression model, which showed a significant improvement from the best simple linear regression model. An interpretation of the fitted multiple variable model and its coefficients, followed by additional analysis using confidence intervals, allowed us to answer the question posed regarding the politician's statement.

This analysis further showcases the utility of the `lm`, `summary`, `cor`, and `plot` functions by demonstrating their usage in the multiple variable case. In addition to expanding the reader's knowledge of previously used functions, the `corrplot` function from the library of the same name and the `confint` function were used to convey the concepts presented earlier in the chapter.

Generally speaking, multiple regression tools can be used in a variety of real-world applications. This case study demonstrates their utility in solving problems in real estate and investment strategies. Solving these kinds of problems is important, as finding the drivers behind price and other monetary variables allows businesses to optimize their profits.

112 5 Multiple Regression

Problems

1. **Automotive Tire Sales Simple and Multiple Regression**
   For this example, we use a subset of the tire sales data set. The data set has several variables and is used to calculate the sales of tires at 500 different stores. The variables of the data set are as follows:

   - Sales—the number of tires sold at each store (in thousands)
   - CompPrice—the local competitor's price for the same or similar tire in USD
   - Income—the average annual income in thousands of USD within a 5 mile radius of the store
   - Ads—the number of ads from a particular store
   - Cars—the average number of cars in a household within a 5 mile radius
   - Price—the sales price of the tire offered in USD
   - SellerType—one of three possible categories: supercenter, automotive, or dealer
   - District—one of the four possible categories denoting the district of the store: north, south, east, or west
   - Age—the average age of customers from each store

   A subset of the data can be accessed using the R code below:

```
df = read.csv("TireSales.csv")
sub = subset(df, select = c('Sales', 'CompPrice', 'Income',
                            'Ads', 'Cars', 'Price', 'Age'))
```

   The data subset sub provides the variables required for the questions below. Using the sub dataframe, do the following:

   a. Get a summary of the dataframe.
   b. For each variable in the dataframe, fit a simple linear regression model to predict Sales.
   c. In which of these models is there a statistically significant relationship between predictor and response? **Hint: the *p*-value is available within the coefficients for each summary. It may be advantageous to create a summary object and access these coefficients directly.**
   d. Fit a multiple regression model predicting Sales as a function of all of the predictors. For which predictors can you reject the null hypothesis $H_0 : \beta_j = 0$?
   e. How do your results from the simple linear regression models compare to those from the multiple regression model?

2. Automotive Tire Sales Correlations

A subset of the data can be accessed using the R code below:

```
df = read.csv("TireSales.csv")
sub = subset(df, select = c('Sales', 'CompPrice', 'Income',
                            'Ads', 'Cars', 'Price', 'Age'))
```

The data subset sub provides the variables required for the questions below. Using the sub dataframe, do the following:

a. Find a correlation plot between the variables within the data subset (Sales, CompPrice, Income, Ads, Population, Price).

b. From the correlation plot, which predictor variable is sales most correlated with?

c. Generate a scatterplot matrix from the variables within the data subset (Sales, CompPrice, Income, Advertising, Population, Price).

d. Which variable relationship appears to be the strongest? Compare your answer with your result from part b.

3. Automotive Tire Sales Modeling

Using the data from TireSales.csv, answer the following questions:

a. Fit a linear model to predict the response variable (Sales) from the variables: CompPrice, Income, Ads, Cars, Price, and Age.

b. Display a summary of the model given from a and clearly state the regression equation.

4. Automotive Tire Sales by Store Type

Using the data from TireSales.csv, answer the following questions:

a. Construct boxplots of the number of sales by store type.

b. It is believed that sales are significantly different across the different store types. Can you refute this claim using an F-test with a significance level of $\alpha = 0.05$? Clearly state the null and alternative hypotheses.

5. Automotive Tire Sales by District

Using the data from TireSales.csv, answer the following questions:

a. Construct boxplots of the number of sales by store type.

b. It is believed that sales are significantly different across the different districts. Can you refute this claim using an F-test with a significance level of $\alpha = 0.05$? Clearly state the null and alternative hypotheses.

Chapter 6
Estimation Intervals and Analysis of Variance

An approximate answer to the right problem is worth a good deal more than an exact answer to an approximate problem.
—*John Tukey*

6.1 Introduction

In some cases, a point estimate alone proves insufficient and requires a confidence interval. Hence, we expand our coverage to explore confidence intervals for the mean response and prediction about the predicted value of the response. Chapters 2 and 3 introduced the fundamental concepts behind sum of squares and the explained and unexplained components. Here, we look at these fundamental concepts more in depth and in the process provide additional analysis techniques.

We begin with a discussion of confidence intervals of the mean response for simple linear regression and then cover prediction intervals for individual values. These intervals of estimation provide different types of estimates, which we in turn explore with an application of housing prices. We then discuss analysis of variance for regression and expand on the house price application. In the final discussion, we present a case study from human resources and solve it in detail. The case study involves finding both prediction and confidence intervals to find interval estimates for both a mean response and an individual response. This case study makes use of previously learned R functions with the aforementioned concepts and shows the relevant source code.

6.2 Expected Value

The expected value or expectation refers to the mean value. As the name implies, the expected value provides the long-term value that we would expect in a given scenario. For instance, if one were to flip a fair coin many times, one would expect an equal number of heads and tails. Therefore, the expected number of heads from many flips of a fair coin would be one half of the flips. It follows that the expectation of one flip is 50% heads. Mathematically, if we let X represent the number of heads, then the expected value for the number of heads is written as $E[X]$. In regression analysis, we use the conditional expectation to find the expected value of Y given X, written as $E[Y|X]$. In fact, making predictions using fitted regression models can be thought of as calculating conditional expectations from known data.

The expected value of the simple linear regression model is written as

$$E[Y|X] = \beta_0 + \beta_1 X, \tag{6.1}$$

and the expected value of the multiple regression model is written as

$$E[Y|X_1, X_2, \ldots, X_p] = \beta_0 + \beta_1 X_1 + \beta_2 X_2 + \ldots + \beta_p X_p. \tag{6.2}$$

Notice that Eqs. (6.1) and (6.2) do not include the error term (ε) since the error has a mean of 0 ($E[\varepsilon] = 0$) as assumed in the assumptions of linear regression. The actual value of $E[Y|X_1, X_2, \ldots, X_p]$ is in fact unknown, but can be estimated as $\hat{Y}$.

6.3 Confidence Interval

As mentioned in the previous section, the expected value of Y given X is synonymous with the mean of Y given X. A confidence interval can be constructed about this mean and, since its estimate is $\hat{Y}$ corresponding to a value of X, the best estimate of Y is $\hat{Y}$.

Assuming the simple linear regression case and using lower case x to denote a particular X value of interest, the standard error of the mean is calculated to be

$$SE_{CI} = s\sqrt{\frac{1}{n} + \frac{(x - \bar{x})^2}{\sum_{i=1}^{n}(x_i - \bar{x})^2}}. \tag{6.3}$$

Using this standard error, a $(1 - \alpha) \times 100\%$ confidence interval for $E[Y|X]$, the mean response of Y for a specified value of X, is given by the interval:

$$\hat{y} \pm t_{\alpha/2} SE_{CI}, \tag{6.4}$$

where $t_{\alpha/2}$ is the critical value with $n - 2$ degrees of freedom for the simple linear regression case.

6.4 House Prices Application: Confidence Interval

Charles is interested in purchasing a house in Dallas and wants to purchase a house that is roughly 2500 square feet. He collects the size and sale price of 12 houses in a particular Dallas neighborhood. The data shown here reflect the size (in square feet) and the sale price (in thousands of dollars). Construct a 95% confidence interval about the predicted mean house price for all houses with a size of 2500 square feet.

Use the data from Table 6.1 to find the following.

(a) Calculate the mean of X ($\bar{x}$).
(b) Calculate the SS_{xx}.
(c) Find the least squares line.
(d) Calculate the $RMSE$.
(e) Predict the house price of a house with 2500 square feet.
(f) Calculate the critical value $t_{\alpha/2}$ for $\alpha = 0.05$.
(g) Find the standard error (SE_{CI}).
(h) Calculate the upper and lower bounds of the confidence interval.
(i) Provide an interpretation of the confidence interval.

Solution

(a) The mean is easily calculated to be $\bar{x} = 2375$. This operation can be done by hand or using R as follows.

```
X = c(3500, 2400, 4900, 1900, 1200, 1600, 1450, 1550, 1600,
      1750, 2850, 3800)
mean(X)
```

```
## [1] 2375
```

(b) From formula (2.5) in Chap. 2, we have

$$SS_{xx} = \sum_{i=1}^{n} x_i^2 - \left(\sum_{i=1}^{n} x_i\right)^2 / n$$

$$SS_{xx} = 82320000 - (28500)^2 / 12$$

$$SS_{xx} = 14,632,500.$$

Table 6.1 House price data

Variable												
Size (X)	3500	2400	4900	1900	1200	1600	1450	1550	1600	1750	2850	3800
Price (Y)	588	490	675	425	350	412	385	405	420	418	509	550

```
SSxx = sum(X^2)-sum(X)^2/12
SSxx
```

    ## [1] 14632500

(c) Using the methods from Chap. 3, we calculate the regression equation:

$$\hat{Y} = 275.082 + 0.081615X$$

We saw in the stock beta case study from Chap. 4 that the coefficients can be calculated from the regression summary as follows. Here the regression coefficients are within the model object `reg`.

```
X = c(3500, 2400, 4900, 1900, 1200, 1600, 1450, 1550, 1600,
      1750, 2850, 3800)
Y = c(588, 490, 675, 425, 350, 412, 385, 405, 420, 418, 509,
      550)
reg = lm(Y ~ X)
reg$coefficients
```

    ##   (Intercept)   X
    ## 275.08209465   0.08161456

(d) From formula (4.4) in Chap. 4, we have

$$s = \sqrt{\frac{SSE}{n-2}}$$

$$RMSE = 17.9799.$$

```
SSE = sum(reg$residuals^2)
SSE
RMSE = sqrt(SSE/(12-2))
RMSE
```

    ## [1] 3232.773

    ## [1] 17.97991

(e) Noting that $\hat{y}$ is calculated from $x = 2500$, the regression equation from part c allows us to calculate

$$\hat{y} = 275.08209465 + 0.08161456(2500) = 479.1185.$$

```
predict(reg, data.frame(X = 2500))
```

```
##           1
## 479.1185
```

(f) The qt function in R allows us to calculate the critical value ($t_{\alpha/2}$) as discussed in Sect. 4.6.1. Since the critical value is such that 2.5% is in each tale, we want the t-value where 97.5% of the area is to the left which leads us to inputting 0.975 as the first argument of the qt function. The second argument is the degrees of freedom $n - 2$. Using the command line in R we have:

```
qt(0.975,10)
```

```
## [1] 2.228139
```

which can be written as $t_{\alpha/2} = t_{0.025} = 2.228$ with 10 degrees of freedom.

(g) The standard error for $x = 2500$ can be obtained by Eq. (6.3):

$$SE_{CI} = s\sqrt{\frac{1}{n} + \frac{(x - \bar{x})^2}{SS_{xx}}}$$

$$SE_{CI} = 17.9799\sqrt{\frac{1}{12} + \frac{(2500 - 2375)^2}{14632500}}$$

$$SE_{CI} = 5.223528.$$

Calculating this value in R is done using the basic operators and the sqrt function.

```
SE_ci = 17.9799*sqrt(1/12+(2500-2375)^2/14632500)
SE_ci
```

```
## [1] 5.223499
```

(h) Using the standard error, the confidence interval is

$$\hat{y} \pm t_{\alpha/2}SE_{CI}$$

which gives a lower bound of

$$479.1185 - 2.228139(5.223499) = 467.4798,$$

and an upper bound of

$$479.1185 + 2.228139(5.223499) = 490.7572.$$

We can find these bounds using R by using the `predict` function and specifying a dataframe with the X value of interest and the `interval` argument specified as a "confidence" interval.

```
predict(reg, data.frame(X=2500), interval = "confidence")
```

```
##          fit      lwr      upr
## 1  479.1185 467.4798 490.7572
```

This output matches the lower and upper bounds of our calculation specified above. These values are 467.4798 and 490.7572 respectively.

(i) The interpretation is such that we are 95% confident that the mean price for houses of 2500 square feet is between \$467,480 and \$490,757.

6.5   Prediction Interval

In many cases, a prediction interval for a particular value seems more desirable. Confidence intervals about a mean response yield the standard error based upon a mean, and therefore, have smaller standard error. As with confidence intervals, prediction intervals are centered about the predicted value of y. Prediction intervals are calculated when we wish to know the range under which $(1 - \alpha) \times 100\%$ of the predictions lie.

Note the subtle difference between the standard error of the confidence interval in Eq. (6.3) and the standard error of the prediction interval:

$$SE_{PI} = s\sqrt{1 + \frac{1}{n} + \frac{(x - \bar{x})^2}{\sum_{i=1}^{n}(x_i - \bar{x})^2}}. \tag{6.5}$$

Similar to that of the confidence interval, a $(1 - \alpha) \times 100\%$ prediction interval for an individual response about $\hat{y}$, for a specified value of $X = x$, is given by

$$\hat{y} \pm t_{\alpha/2}SE_{PI}, \tag{6.6}$$

where x is the given value of the predictor variable, n is the number of observations, and $t_{\alpha/2}$ is the critical value with $n - 2$ degrees of freedom.

In the application below, we will show and discuss a comparison between the prediction and confidence intervals.

6.6 House Price Application: Prediction Interval

Construct a 95% prediction interval about the predicted house price for all houses with a size of 2500 square feet.

Use the data from Table 6.1 to find the following.

(a) Calculate the upper and lower bounds of the prediction interval.
(b) Provide an interpretation of the prediction interval.
(c) Explicitly state the difference between the interpretation of the prediction interval and the confidence interval.
(d) Indicate which interval has a larger range.

Solution

(a) Recall from the previous application that $SS_{xx} = 14{,}632{,}500$, $s = 17.9799$, $t_{0.025} = 2.228139$, and $\hat{y} = 479.1185$. Using the aforementioned calculations, we find the standard error of the prediction interval at $x = 2500$:

$$SE_{PI} = s\sqrt{1 + \frac{1}{n} + \frac{(x - \bar{x})^2}{SS_{xx}}}$$

$$SE_{PI} = 17.9799\sqrt{1 + \frac{1}{12} + \frac{(2500 - 2375)^2}{14632500}}.$$

$$SE_{PI} = 18.72329$$

In R, the calculation can also be easily found doing the following.

```
SE_pi = 17.9799*sqrt(1+1/12+(2500-2375)^2/14632500)
SE_pi
```

```
## [1] 18.72329
```

Using the standard error, the prediction interval is

$$\hat{y} \pm t_{\alpha/2} SE_{PI}$$

which gives a lower bound of

$$479.1185 - 2.228139(18.72329) = 437.4004,$$

and an upper bound of

$$479.1185 + 2.228139(18.72329) = 520.8366.$$

We can find these bounds using R by using the `predict` function and specifying a dataframe with the X value of interest and the `interval` argument specified as a "prediction" interval.

```
predict(reg, data.frame(X = 2500), interval = "prediction")
```

```
##         fit      lwr      upr
## 1 479.1185 437.4004 520.8366
```

This output matches the lower and upper bounds of our calculation specified above. These values are 437.4004 and 520.8366, respectively.

(b) The interpretation is such that we are 95% confident that house price for a random house with a size of 2500 square feet is between \$437,400 and \$520,837.

(c) The result from this application represents the interval of a particular house price whereas the previous application represents that of a mean house price.

(d) The prediction interval is a larger range interval than the confidence interval.

6.7   Confidence Intervals verse Prediction Intervals

In the previous applications, both confidence and prediction intervals were calculated. However, note that confidence intervals give a range for $E[y|x]$ which is the mean response of Y and prediction intervals give a range for values of y. In both cases, the best guess of a particular value of y given a known x is $\hat{y}$, and therefore, both intervals are centered at $\hat{y}$. Since the mean has less variation than a single value, the standard error of $E[Y|X]$ is smaller than the standard error of y. Furthermore, the prediction interval will have a larger range than that of a confidence interval, which is attributed to the standard errors (Eqs. 6.3 and 6.5).

6.8   Analysis of Variance

Analysis of Variance (ANOVA), developed by statistician and evolutionary biologist Ronald Fisher, is a collection of statistical models used to analyze the differences among group means and their associated procedures (such as "variation" among and between groups).

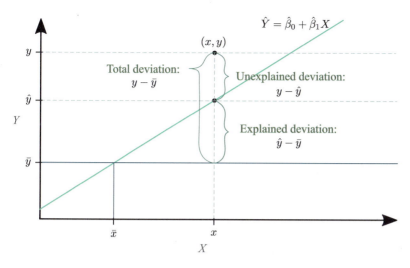

Fig. 6.1 Explained vs. unexplained deviation

By analyzing the variation, the least squares regression equation becomes a natural result. The components of the variation in regression modeling are represented by the sum of squares of the total, regression, and the error: SST, SSR, and SSE, respectively. Recall from Chap. 3 the relationship between the sum of squares in Eq. (3.11):

$$SST = SSR + SSE.$$

This equation holds not only for the simple linear regression case but also for the multiple regression case. Figure 6.1 provides an intuitive depiction between the total, unexplained, and explained deviations. These deviations represent the components of the corresponding sum of squares values.

6.8.1   Mean of Squares Due to Regression

While the SSR is variation, it is preferred that this variation be as large as possible since SSR represents the amount of variation "explained" by the regression model. It follows that the larger values of the Mean of Squares due to Regression (MSR) indicate large amounts of explained variation. The MSR represents the mean variation explained by each predictor. From the SSR and the degrees of freedom, the Mean of Squares due to Regression (MSR) can be calculated as

$$MSR = SSR/p. \tag{6.7}$$

The degrees of freedom for the SSR corresponds to the number of predictors (p) in the model. For the simple linear regression model, there is one degree of freedom since we have only one predictor variable (X). In the case of a model with $p = 10$ predictors, there are 10 degrees of freedom. The MSR, the SSR, and corresponding degrees of freedom are shown in the ANOVA table (Table 6.2).

6.8.2 Mean Squared Error

The degrees of freedom for the SSE corresponds to the number of observations (n) minus the predictors (p) minus 1 or

$$n - p - 1.$$

In the simple linear regression case, there are $n - 2$ degrees of freedom. For the model with 20 observations and 7 predictors, the degrees of freedom of the SSE would be $20 - 7 - 1 = 12$.

The previous chapters were devoted to simple linear regression and resulted in $n - 2$ used as the degrees of freedom, with the MSE found by dividing by $n - 2$. Generally, the MSE can be found by

$$MSE = \frac{SSE}{n - p - 1}. \tag{6.8}$$

Both SSE and MSE are considered measures of error; therefore, we desire them to be as small as possible. The MSE, SSE, and corresponding degrees of freedom are shown in the ANOVA table (Table 6.2).

6.8.3 The F Statistic

Taking the ratio of the MSR to MSE results in the F-statistic:

$$F = \frac{MSR}{MSE}. \tag{6.9}$$

Since the degrees of freedom of the MSR is p and the degrees of freedom of the MSE is $n - p - 1$, the F-statistic has p degrees of freedom in its numerator and

Table 6.2 The ANOVA table

Variation source	DF	SS	MS	F	p-value
Regression	p	SSR	$MSR = \frac{SSR}{p}$	$F = \frac{MSR}{MSE}$	p-value (F)
Error	$n - p - 1$	SSE	$MSE = \frac{SSE}{n-p-1}$		
Total	$n - 1$	SST			

Table 6.3 The ANOVA table for simple linear regression

Variation source	DF	SS	MS	F	p-value
Regression	1	SSR	$MSR = \frac{SSR}{1}$	$F = \frac{MSR}{MSE}$	p-value (F)
Error	$n-2$	SSE	$MSE = \frac{SSE}{n-2}$		
Total	$n-1$	SST			

$n - p - 1$ degrees of freedom in its denominator. Noting the ratio of MSR to MSE, a larger F-statistic results in lower p-values, indicating overall significance.

6.9 ANOVA Table

From the ANOVA components in the previous section, a table can be created to summarize the F-statistic. Specifically, the MSR, MSE, and their underlying components can be shown in an ANOVA table (Table 6.2).

Considering the simple linear regression case, the ANOVA table can be simplified such that $p = 1$. Filling in $p = 1$ for Table 6.2 results in Table 6.3. In this table,

$$MSR = \frac{SSR}{1} = SSR$$

and

$$MSE = \frac{SSE}{n - 2}$$

as discussed in Chap. 4.

Using the anova function in base R, the components of ANOVA are easily returned from the fitted regression model. For instance, if the regression model object is named reg, then the anova function can be used in the following manner:

```
anova(reg)
```

We demonstrate the utility of this function in the following application.

6.10 House Price Application: ANOVA Table

Calculate the values in the ANOVA table for the data from Table 6.1. In particular, do the following to construct the ANOVA table.

(a) Calculate the SSR.
(b) Calculate the SST.

(c) Calculate the MSR.
(d) Calculate the F-statistic.
(e) Use R to find the p-value of the F-statistic.
(f) Use the calculated information from the previous parts to construct the ANOVA table for the data.
(g) Verify your table using the anova function in R.

Solution

(a) Using Eq. (3.17), the SSR is

$$SSR = \sum_{i=1}^{n}(\hat{y} - \bar{y})^2$$

$$SSR = 97466.14.$$

While we could easily find the $\hat{y}$ values using some simple calculation, these values are given to us within the regression object reg accessed as reg$fitted.values. Using the vector already calculated for us, we use the sum function and pay careful attention to the orders of operation by adding in extra parentheses.

```
SSR = sum((reg$fitted.values - mean(Y))^2)
SSR
```

```
## [1] 97466.14
```

(b) Equation (3.16) is

$$SST = \sum_{i=1}^{n}(y_i - \bar{y})^2$$

$$SST = 100698.9$$

Using R, the calculation is done using the formula here.

```
SST = sum((Y - mean(Y))^2)
SST
```

```
## [1] 100698.9
```

Equivalently, the SST can be found with Eq. (3.19):

$$SST = SSR + SSE$$

which yields the same result (using $SSE = 3232.773$).

(c) For the MSR, we have $MSR = SSR/1 = 97{,}466.14$, since $p = 1$.

(d) Combining Eqs. (6.9) and (6.8), the F-statistic is

$$F = \frac{MSR}{SSE/(n-2)}$$

$$F = \frac{97{,}466.14}{3232.773/10}$$

$$F = 301.4939.$$

(e) The pf function in R allows us to calculate the p-value of the F-statistic as discussed in Sect. 4.14. Using the pf function with $F = 301.4939$, $v_n = 1$, and $v_d = 10$, we have:

```
1 - pf(301.4939, 1, 10)
```

```
## [1] 8.50638e-09
```

(f) Filling in the calculated values, we obtain Table 6.4.

(g) Passing the regression object to the anova function, we have:

```
X = c(3500, 2400, 4900, 1900, 1200, 1600, 1450, 1550, 1600,
      1750, 2850, 3800)
Y = c(588, 490, 675, 425, 350, 412, 385, 405, 420, 418, 509,
      550)
reg = lm(Y ~ X)
anova(reg)
```

```
## Analysis of Variance Table
##
## Response: Y
##             Df Sum Sq Mean Sq F value    Pr(>F)
## X            1  97466   97466  301.49 8.506e-09 ***
## Residuals   10   3233     323
```

Table 6.4 The ANOVA table for the house price data

Variation source	DF	SS	MS	F	p-value
Regression	1	97,466	97,466	301.49	8.506e−09
Error	10	3232.8	323.28		
Total	11	100,698.9			

```
## ---
## Signif. codes:
## 0 '***' 0.001 '**' 0.01 '*' 0.05 '.' 0.1 ' ' 1
```

While the labels differ slightly, the information in the R table is consistent with the table created in part f. We see a slight rounding error in the p-value, but this error is negligible and does not affect the test results with $\alpha = 0.05$. The anova function does, however, leave out the total degrees of freedom and the SST, which are unnecessary in finding the F-statistic.

6.11 Generalized F Statistic

Arguably, the most important F-test for multiple linear regression consists of the test for overall significance. However, this test can be modified from the hypothesis tests in Chap. 4 to assess differences between models. Recall from Chap. 4 the null and alternative hypotheses for an F-test:

$$H_0 : \beta_1 = \beta_2 = \ldots = \beta_p = 0$$

$$H_1 : \text{At least one coefficient is nonzero.}$$

The null and alternative hypotheses can be thought of in terms of the models that result. For example, the null hypothesis with $\beta_1 = \beta_2 = \ldots = \beta_p = 0$ would give us the following reduced model:

$$H_0 : Y = \beta_0 + \varepsilon$$

and the alternative hypothesis would result in the full model:

$$H_1 : Y = \beta_0 + \beta_1 X_1 + \beta_2 X_2 + \cdots + \beta_p X_p + \varepsilon.$$

The F-statistic was previously calculated to be

$$F = \frac{SSR/p}{MSE}.$$

Recall that the SSR is the difference between the sum of squares total (SST) and the sum of squared error in the full model (SSE). The F-statistic can therefore be written as

$$F = \frac{(SST - SSE)/p}{MSE}. \tag{6.10}$$

The test statistic, as calculated in Eq. (6.10), represents the difference between the SST and the SSE. The SST is the amount of error that would result from assuming the null hypothesis that $\beta_1 = \beta_2 = \ldots = \beta_p = 0$, whereas, the SSE is the error that results from using the alternative hypothesis.

Now consider the reduced model (null hypothesis model) to contain one less predictor variable than the full model (alternative hypothesis model). If the full model contains p predictors, then the statistical test to compare the models with and without X_p would be:

$$H_0 : Y = \beta_0 + \beta_1 X_1 + \beta_2 X_2 + \cdots + \beta_{p-1} X_{p-1} + \varepsilon$$

$$H_1 : Y = \beta_0 + \beta_1 X_1 + \beta_2 X_2 + \cdots + \beta_p X_p + \varepsilon.$$

The numerator of the F-statistic, for the test mentioned here, is the difference between the sum of squared error in the reduced model and the sum of the squared error of the full model divided by the number of predictors. Here we denote the sum of the squared error in the reduced model to be $SSE(X_1, X_2, \ldots, X_{p-1})$. Since the difference in the number of predictors is equal to one, the degrees of freedom in the numerator are equal to one, resulting in an F-statistic of

$$F = \frac{SSE(X_1, X_2, \ldots, X_{p-1}) - SSE}{MSE}. \tag{6.11}$$

In general, it is possible to generalize the F-statistic to compare models with any number of predictor variables as long as the reduced model is a subset of the full model. For simplicity, the Eq. (6.11) only considers the comparison of models with a difference of one predictor variable.

6.12   Case Study: Employee Retention Modeling

6.12.1   Problem Statement

Managers at modern companies work hard to ensure that their employees are satisfied with their current jobs. Several corporate problems stem from employees' lack of satisfaction in the workplace, including costly restaffing fees or a lacking end product due to poor quality of work or productivity. Companies hence investigate when a disproportionate number of employees churn from their current roles.

Angelo, a hiring manager at Royalty Cruise Lines, carries out the task of predicting future employee retention in new applicants. He collects data from several current employees and, based on the data, creates a regression model. In particular, he prepares to predict the number of years an employee will stay with Royalty using the number of years that they worked at a previous job, education level, starting salary, and a dummy variable (a more thorough coverage of dummy variables will follow in the next chapter) that states if they would be hired in branch

A of the company. Angelo has a particular interest in a few employees who will be working with him on an important project.

Angelo has to decide among seven job candidates. If he would like the employee with the longest retention time, whom should he choose? For this role, Angelo would prefer that the candidate remains at his job for at least five years. Calculate a 95% prediction interval for this candidate's retention.

Angelo also faces the challenge of whether to hire applicants from competing staffing and recruiting companies Gravity AP or Ascend. The applicants from these companies generally have different characteristics due to differing recruiting strategies.

6.12.2   Data Description

The data consists of 100 former employees followed throughout their career at Royalty. The variables for each employee are:

- Retention—The number of years that the employee worked at Royalty.
- BranchA—Denotes whether employee worked in branch A during their work career (1 if they worked in branch A; 0 otherwise).
- Education—The education level in years.
- Experience—The number of years of previous experience before joining Royalty.

Here we load in the data and use the head and summary functions to peruse the data.

```
df = read.csv("HR_retention.csv")
head(df)
```

```
##     Retention BranchA Education Experience
## 1   6.9680996       0        19  12.345578
## 2   0.1956849       0        15   7.669670
## 3   5.3079883       0        19   8.152030
## 4   6.9384618       0        19  10.139741
## 5   1.9794614       0        17   6.608843
## 6  11.1552751       0        13  11.730156
```

```
summary(df)
```

```
##    Retention           BranchA          Education
## Min.   : 0.1957    Min.   :0.0      Min.   :12.00
## 1st Qu.: 2.3841    1st Qu.:0.0      1st Qu.:15.00
## Median : 5.5609    Median :0.5      Median :16.00
## Mean   : 6.4017    Mean   :0.5      Mean   :16.05
```

Table 6.5 Recent hires data

BranchA	Experience	Education
1	0	12
0	10	16
0	0	16
1	10	16
0	15	18
1	10	16
1	10	18

```
##   3rd Qu.: 9.8388    3rd Qu.:1.0    3rd Qu.:17.00
##   Max.    :21.3770    Max.    :1.0    Max.    :21.00
##      Experience
##   Min.    : 4.959
##   1st Qu.: 7.400
##   Median : 9.389
##   Mean    : 9.863
##   3rd Qu.:11.760
##   Max.    :17.976
```

The employees that Angelo recently hired have the characteristics given in Table 6.5.

The data set can be arranged into a dataframe by creating vectors for each variable using the c function, and then using the data.frame function.

```
BranchA = c(1, 0, 0, 1, 0, 1, 1)
Experience = c(0, 10, 0, 10, 15, 10, 10)
Education = c(12, 16, 16, 16, 18, 16, 18)
df_new = data.frame(BranchA, Experience, Education)
```

Running the above code provides the information from the new hires into the dataframe df_new.

6.12.3 Multiple Regression Model

As discussed previously, Angelo's firm would like us to predict the average retention based upon all of the other variables in the data. Therefore, our model will be a multiple regression model:

$$Y = \beta_0 + \beta_1 X_1 + \beta_2 X_2 + \beta_3 X_3 + \varepsilon,$$

where Y, X_1, X_2, and X_3 variables are Retention, BranchA, Education, and Experience, respectively. The remaining component in the model, ε, refers to the error.

The regression equation, as determined by the summary above, is given by:

$$\hat{Y} = -7.93599 + 1.43316X_1 - 0.01740X_2 + 1.40928X_3.$$

Substituting in for the full variable names we have:

$$Retention = -7.93599 + 1.43316 \times BranchA - 0.01740 \times Education$$
$$+ 1.40928 \times Experience.$$

The regression model fit is given below:

```
reg = lm(Retention ~ BranchA + Education + Experience,
         data = df)
summary(reg)
```

```
##
## Call:
## lm(formula = Retention ~ BranchA + Education + Experience,
##     data = df)
##
## Residuals:
##     Min      1Q  Median      3Q     Max
## -4.3464 -1.2415 -0.1563  1.2840  4.8966
##
## Coefficients:
##              Estimate Std. Error t value Pr(>|t|)
## (Intercept) -7.93599    1.82783  -4.342 3.5e-05 ***
## BranchA      1.43316    0.40929   3.502 0.000704 ***
## Education   -0.01740    0.11021  -0.158 0.874915
## Experience   1.40928    0.07165  19.669  < 2e-16 ***
## ---
## Signif. codes:
## 0 '***' 0.001 '**' 0.01 '*' 0.05 '.' 0.1 ' ' 1
##
## Residual standard error: 2.014 on 96 degrees of freedom
## Multiple R-squared:  0.8219, Adjusted R-squared:  0.8164
## F-statistic: 147.7 on 3 and 96 DF,  p-value: < 2.2e-16
```

Model Interpretation

Since the F-statistic has a very low p-value in the summary, the model proves statistically significant. The high values of the R^2 and R_a^2 indicate the model is fit well. The most significant predictor variable based on the p-values consists of

Experience, followed by `BranchA`. The `Education` variable does not appear to be significant since its p-value is well above 0.05.

Coefficient Interpretation

Fitting a regression model tests each coefficient estimate against a null value of zero. The coefficients can be interpreted as follows:

- Intercept—The intercept represents the retention when all the variables are 0. The p-value is significant since it falls below 0.05. In this case, the retention is below 0 which does not make sense.
- BranchA—Since the p-value indicates significance, the interpretation is:

  The retention will increase by 1.433 if they are to work in Branch A holding experience and education constant.

- Education—Since the p-value indicates the variable is insignificant, the coefficient for education should not be interpreted.
- Experience—The p-value indicates this variable is significant. The interpretation is:

  For every 1 year of previous experience, the retention is expected to increase by 1.409 years, holding Branch A and education constant.

ANOVA Table

Within the ANOVA table, we should note several numerical computations which are the fundamental building blocks of the linear regression equation. We easily calculate these values using the `anova` function in base R.

```
anova(reg)
```

```
## Analysis of Variance Table
##
## Response: Retention
##             Df  Sum Sq Mean Sq  F value      Pr(>F)
## BranchA      1  200.54  200.54  49.4271  3.005e-10 ***
## Education    1   27.53   27.53   6.7862    0.01065 *
## Experience   1 1569.69 1569.69 386.8763  < 2.2e-16 ***
## Residuals   96  389.51    4.06
## ---
## Signif. codes:
## 0 '***' 0.001 '**' 0.01 '*' 0.05 '.' 0.1 ' ' 1
```

Note that the order of the variables matters. The F-statistic for BranchA is calculated using the sum of squares of 200.54. This sum of squares represents the variation explained by using BranchA as a prediction. The sum of squares of Education represents the benefit of using the BranchA variable and the Education predictor over using only the BranchA variable.

$$F = \frac{SSE(X_1, X_2) - SSE}{MSE}.$$

Therefore, one should note that $SSE(X_1, X_2) = 1569.69 + 389.51$ represents the reduced sum of squared error since 1569.69 is the variation explained through including experience in the model and 389.51 is the SSE that would still be included. The F-statistic for a model without experience would be:

```
F = (1569.69)/(389.51/(100-4))
F
```

```
## [1] 386.8713
```

Notice that this F value corresponds to the F value for experience in the R table above.

From the output above, we can calculate the SST by summing up all of the sum of squares values. Particularly, we have:

```
SST = 200.54+27.53+1569.69+389.51
SST
```

```
## [1] 2187.27
```

One should further note that the value of R^2 can be calculated using these sum of squares values:

$$R^2 = SSR/SST = 1 - SSE/SST$$

```
Rsquare = 1-389.51/SST
Rsquare
```

```
## [1] 0.8219196
```

Furthermore, R_a^2 can be calculated as:

$$R_a^2 = 1 - (1 - R^2)\frac{n-1}{n-p-1}$$

```
AdjRsquare = 1 - (1-Rsquare)*(100-1)/(100-3-1)
AdjRsquare
```

```
## [1] 0.8163545
```

The values of R^2 and R_a^2 can be verified by viewing the regression summary.

6.12.4 Predictions

Predict the retention of an employee who will be working in branch A with no work experience and 12 years of education.

$$Retention = -7.93599 + 1.43316 \times 1 - 0.01740 \times 12 + 1.40928 \times 0$$

$$= -6.7116.$$

Making predictions in R is straight-forward using the `predict` command. The first argument is the regression object and the second argument is a dataframe. Note, the variable names in the dataframe must reflect those of the predictor variables used within the `lm` function.

```
predict(reg, data.frame(BranchA = 1, Experience = 0,
                        Education = 12))
```

```
##          1
## -6.711566
```

Negative retention values are impossible, and therefore, the result of this retention calculation does not make sense. Based on the model above, we would predict the retention to be 0. Further note that this prediction represents an extrapolation since our original data set has minimum years of experience of almost 5 years.

Recall that the dataframe of new hires defined previously, df_new, can be predicted.

```
predict(reg, df_new)
```

```
##          1          2          3          4          5
## -6.711566   5.878533  -8.214309   7.311696  12.890165
##          6          7
##   7.311696   7.276906
```

Based on this model, Applicant 5 who has 15 years of experience, is predicted to have the highest retention in the group.

Prediction Interval

The `predict` function can also be modified to return the limits of the prediction or confidence interval by specifying the `interval` option within the function. Calculate 95% and 99% prediction intervals for applicant 5. The 95% prediction interval serves as the default, given by the `predict` function:

```
predict(reg, df_new[5,], interval = "prediction")
```

```
##          fit      lwr      upr
## 5 12.89016 8.763318 17.01701
```

The `predict` function may be easily modified to return a prediction interval with a different α using the `level` option within the `predict` function. A 99% prediction interval for Applicant 5 can be found through

```
predict(reg, df_new[5,], interval = "prediction", level = 0.99)
```

```
##          fit      lwr     upr
## 5 12.89016 7.426434 18.3539
```

Observing a 99% prediction interval for Applicant 5, we are very confident that he or she will remain at Royalty for more than 5 years as desired.

Confidence Intervals

Ascend mentions that their typical applicants have the properties given in Table 6.6. Whereas applicants from Gravity AP have the properties given in Table 6.7. Find 95% confidence intervals for each set of applicants.

```
df_ascend = data.frame(BranchA=1, Experience=10, Education=20)
predict(reg, df_ascend, interval = "confidence")
```

Table 6.6 Ascend data

BranchA	Experience	Education
1	10	20

Table 6.7 Gravity AP data

BranchA	Experience	Education
1	7	16

```
##          fit      lwr      upr
## 1 7.242116 6.196755 8.287477
```

```
df_gravity = data.frame(BranchA = 1, Experience = 7,
                        Education = 16)
predict(reg, df_gravity, interval = "confidence")
```

```
##          fit     lwr      upr
## 1 3.083843 2.34395 3.823737
```

The observed confidence intervals and predictions indicate that hiring Ascend applicants would result in higher retention values.

6.12.5   Case Conclusion

In this analysis, we specified and fit a multiple regression model using R and then interpreted the model and its coefficients. Some additional analysis included the ANOVA table and the calculations of both prediction and confidence intervals for the coefficients. One new hire, in particular, yielded a prediction value that indicated the longest retention. The prediction interval for the retention of this employee was calculated to be sufficient at the $\alpha = 0.01$ level of significance. Both staffing companies had confidence intervals calculated about their mean candidate, showing Ascend as the better fit due to the higher retention values.

Many possible ways to use the results of regression analysis exists. This case study demonstrates how to make use of prediction and confidence intervals and how to calculate ANOVA. While we only introduced a few R commands in this chapter, we built upon knowledge from previous chapters and used many of the previous functions in new ways. When studying regression analysis, it is important to know and understand the various uses of such functions.

Problems

1. **Coral Gables Housing Confidence Interval**
   Jigar is interested in purchasing a house in Coral Gables and wants to get the largest total living area for the price. He collects the size and sale price of 5 houses in a particular Coral Gables neighborhood. The data shown in Table 6.8 reflect the size (in square feet) and the sale price (in thousands of USD).
   Without the use of a computer:

   a. Find the regression equation that predicts price as a function of the house size.

Table 6.8 Coral Gables
house price data

Size	Price
2700	650
3500	875
2600	670
5300	1200
4200	981

   b. Calculate the 95% critical value $t_{\alpha/2}$ for $\alpha = 0.05$ (use R here).
   c. Use $\bar{x} = 3660$, $SS_{xx} = 5{,}052{,}000$, and $RMSE = 26.71$ to find the standard error for a 95% confidence interval (SE_{CI}) for a house price with a size of 2500 square feet.
   d. Calculate the upper and lower bounds of the 95% confidence interval for a house price with a size of 2500 square feet.
   e. Provide an interpretation of the confidence interval.

2. **Coral Gables Housing Prediction Interval**
   Use the Coral Gables housing data from Table 6.8 and do the following without the use of a computer.

   a. Use $\bar{x} = 3660$, $SS_{xx} = 5{,}052{,}000$, and $RMSE = 26.71$ to find the standard error for a 95% prediction interval (SE_{PI}) for a house price with a size of 2500 square feet.
   b. Calculate the upper and lower bounds of the prediction interval for a house price with a size of 2500 square feet.
   c. Provide an interpretation of the prediction interval.
   d. Explicitly state the difference between the interpretation of the prediction interval and the confidence interval.
   e. Indicate which interval has a larger range.

3. **Coral Gables Housing Intervals in R**
   Use the Coral Gables housing data from Table 6.8 and do the following using R.

   a. Print a summary of the regression equation that predicts price as a function of house size.
   b. Use the regression equation to predict the price when the size of the house is 2500 square feet.
   c. Find a 95% confidence interval for the price when the size of the house is 2500 square feet.
   d. Find a 95% prediction interval for the price when the size of the house is 2500 square feet.

4. **Coral Gables Housing ANOVA**
   Using the Coral Gables Housing data from Table 6.8, do the following calculations without the use of a computer.

    a. Calculate the SSR.
    b. Calculate the SST.
    c. Calculate the MSR.
    d. Calculate the F-statistic.

5. **Coral Gables Housing ANOVA in R**
   Using the Coral Gables Housing data from Table 6.8, do the following using R.

    a. Find the p-value of the F-statistic calculated in the previous problem.
    b. Use R to construct the ANOVA table.

6. **Streaming Service Confidence Interval**
   A popular video streaming service relies on paid subscribers. In Table 6.9, we display the revenue (in millions of $) and the number of paid memberships (in millions) for the years 2007 to 2022.
       Without the use of a computer:

    a. Find the regression equation that predicts revenue as a function of the number of memberships.
    b. Use the regression equation to predict the revenue when the memberships is 100 million.
    c. Calculate the critical value $t_{\alpha/2}$ for $\alpha = 0.05$ (use R here).
    d. Find the standard error for a 95% confidence interval (SE_{CI}) for revenue when there are 100 million memberships.
    e. Calculate the upper and lower bounds of the 95% confidence interval centered about 100 million memberships.
    f. Provide an interpretation of the confidence interval.

Table 6.9 Streaming service data

Year	Memberships (X)	Revenue (Y)
2007	1.19	339
2008	5.78	499
2009	6.93	600
2010	7.08	687
2011	9.63	836
2012	12.28	1079
2013	11.10	1604
2014	7.24	1807
2015	23.18	2189
2016	27.03	2752
2017	34.78	3395
2018	38.67	4417
2019	56.84	5845
2020	71.32	7890
2021	88.98	10, 081
2022	101.44	12, 498

7. **Streaming Service Prediction Interval**
   Use the streaming service data from Table 6.9 and do the following without the use of a computer.

   a. Find the standard error for a 95% prediction interval (SE_{PI}) for revenue when there are 100 million memberships.
   b. Calculate the upper and lower bounds of the prediction interval for revenue when there are 100 million memberships.
   c. Provide an interpretation of the prediction interval in part b.
   d. Explicitly state the difference between the interpretation of the prediction interval and the confidence interval.
   e. Indicate which interval has a larger range.

8. **Streaming Service Intervals in R**
   Use the streaming service data from Table 6.9 and do the following using R.

   a. Print a summary of the regression equation that predicts revenue as a function of the number of memberships.
   b. Use the regression equation to predict the revenue when the memberships is 100 million.
   c. Find a 95% confidence interval for the revenue when the memberships is 100 million.
   d. Find a 95% prediction interval for the revenue when the memberships is 100 million.
   e. Find a 95% confidence interval for the slope coefficient.

9. **Streaming Service ANOVA**
   Use the streaming service data from Table 6.9, do the following calculations without the use of a computer.

   a. Calculate the SSR.
   b. Calculate the SST.
   c. Calculate the MSR.
   d. Calculate the SSE.
   e. Calculate the F-statistic.

10. **Streaming Service ANOVA Calculations in R**
    Use the streaming service data from Table 6.9, do the following using R.

    a. Calculate the SSR.
    b. Calculate the SST.
    c. Calculate the MSR.
    d. Calculate the SSE.
    e. Calculate the F-statistic.

11. **Streaming Service ANOVA in R**
    Use the streaming service data from Table 6.9, do the following using R.

    a. Find the p-value of the F-statistic calculated in the previous problem.
    b. Use the anova function in R to construct the ANOVA table.

Chapter 7
Predictor Variable Transformations

If you torture the data enough, nature will always confess.
—*Ronald Coase*

7.1 Introduction

In this chapter, we discuss transformations of predictor variables. One popular transformation consists of dummy variables, which are variables that allow for the effect of categorical variables to be considered in regression modeling. Dummy variables can be used in regression analysis as both predictor and response variables, but we will limit our discussion to predictor variables. Using dummy variables as the response variable is often referred to as classification, which will remain outside of the scope of this book. In previous chapters, we assumed linear models with untransformed predictor variables. Here, we introduce nonlinear transformations of predictor variables, thereby, making the model linear.

First, we discuss categorical variables containing two different categories, and then broaden the discussion to include more than two categories. Applications using salary data are used to reinforce the concepts. Second, curvilinear relationships are introduced, which can be modeled using nonlinear transformations. Third, interactions between predictor variables are explained, followed by a discussion of the general linear model. The final discussion consists of a case study that predicts the number of likes of YouTube videos based upon the characteristics and sentiment of the comments.

© The Author(s), under exclusive license to Springer Nature Switzerland AG 2023
D. P. McGibney, *Applied Linear Regression for Business Analytics with R*,
International Series in Operations Research & Management Science 337,
https://doi.org/10.1007/978-3-031-21480-6_7

7.2   Categorical Variables

Not all data is numeric or quantitative in nature. For instance, a categorical variable refers to a non-numeric variable that can take on one or more possible groupings or categories. Therefore, each observation can be assigned to a particular group. One example of a categorical variable is business type, which can be broken down into two categories: for-profit and not-for-profit. Methods of payment used at a convenience store can also be considered a categorical variable.

It is often the case that categorical data has a relationship to a numeric response variable, which can help predict the response variable. Also, while categorical variables may seem problematic because of their non-numeric characteristic, it is possible to convert categorical variables into quantitative ones. For example, X might represent business type, for a specific company. We can let 0 indicate a not-for-profit business, and 1 indicate a for-profit business. For a particular observation of $X = x$, we specify

$$x = \begin{cases} 0, & \text{if not-for-profit} \\ 1, & \text{if for-profit.} \end{cases}$$

In this case, X is called a dummy or indicator variable. When dummy variables are used in a regression model, the dummy variable coefficients are interpreted in relation to a base level. Here, the dummy variable X denotes the for-profit category and the base level is not-for-profit, since for-profit corresponds to 1 and not-for-profit corresponds to 0. For example, the coefficient for X is 20 indicates that for-profit businesses contribute 20 more to the response variable than the not-for-profit businesses.

Care should be taken in defining dummy variables since their definitions will dictate their interpretations. Dummy variables provide a convenient and easy way to account for categorical data, as we demonstrate in the application below.

7.3   Employee Salary Application: Dummy Variables

Suppose the management of Northrop Grumman want to test whether their employees' annual salaries relate to education level and experience. The education level, highest degree achieved, years of experience, and the annual salary (in thousands of dollars) for each of the sampled 28 employees are shown in the data table above (Table 7.1). The degree variable contains three levels: a Bachelor of Arts (B.A.) degree, a Master of Science (M.S.) degree, and a Master of Science in Business Analytics (M.S.B.A.) degree.

Use the data from Table 7.1 to do the following.

(a) Specify the model to be fit using "Education" and "Experience" variables to predict salary.

Table 7.1 Salary data

Education	Degree	Experience	Salary
Graduate	MSBA	2	85
Graduate	MS	3	74
Graduate	MS	8	100
Graduate	MSBA	0	78
Undergraduate	BA	2	52
Undergraduate	BA	0	60
Graduate	MSBA	0	85
Graduate	MSBA	0	74
Graduate	MSBA	5	124
Graduate	MS	0	73
Graduate	MSBA	2	82
Undergraduate	BA	3	55
Undergraduate	BA	10	115
Undergraduate	BA	0	43
Graduate	MSBA	2	97
Graduate	MS	0	74
Graduate	MS	0	72
Graduate	MSBA	0	80
Undergraduate	BA	5	75
Undergraduate	BA	0	62
Graduate	MSBA	2	90
Graduate	MSBA	3	88
Graduate	MS	8	72
Graduate	MS	0	65
Graduate	MSBA	2	85
Undergraduate	BA	0	54
Undergraduate	BA	20	103
Graduate	MSBA	0	78

(b) Manually code a dummy variable denoting whether or not an employee has a graduate degree. Use 0 to denote if a candidate has an undergraduate degree, and use 1 to denote if a candidate has a graduate degree.

(c) Fit the linear regression model.

(d) Explicitly state the linear regression equation.

(e) Interpret the intercept coefficient.

(f) Interpret the variable coefficients.

(g) Discuss the model fit.

Solution

(a) The multiple linear regression model can be written as

$$Y = \beta_0 + \beta_1 X_1 + \beta_2 X_2 + \varepsilon,$$

where

- Y is the annual salary (in thousands).
- X_1 is the years of experience.
- X_2 is a graduate degree dummy variable (1 for an observation with a graduate degree, 0 otherwise).
- ε is the error.

(b) Using the c function, the degree variable is

```
degree_dummy = c(1, 1, 1, 1, 0, 0, 1, 1, 1, 1, 1, 0, 0, 0, 1, 1,
                 1, 1, 0, 0, 1, 1, 1, 1, 1, 0, 0, 1)
```

(c) The regression model is fit by creating the variables and using the lm function. The model details are then printed using the summary function:

```
experience = c(2, 3, 8, 0, 2, 0, 0, 0, 5, 0, 2, 3, 10, 0, 2, 0,
               0, 0, 5, 0, 2, 3, 8, 0, 2, 0, 20, 0)
salary = c(85, 74, 100, 78, 52, 60, 85, 74, 124, 73, 82, 55,
           115,  43, 97, 74, 72, 80, 75, 62, 90, 88, 72, 65,
           85, 54, 103, 78)
reg = lm(salary ~ experience + degree_dummy)
summary(reg)
```

```
##
## Call:
## lm(formula = salary ~ experience + degree_dummy)
##
## Residuals:
##     Min      1Q  Median      3Q     Max
## -28.554  -6.356   0.082   4.264  32.173
##
## Coefficients:
##                 Estimate Std. Error t value Pr(>|t|)
## (Intercept)      55.8495     4.9308  11.327 2.45e-11 ***
## experience        2.9089     0.5771   5.040 3.36e-05 ***
## degree_dummy     21.4332     5.3113   4.035 0.000452 ***
## ---
## Signif. codes:
## 0 '***' 0.001 '**' 0.01 '*' 0.05 '.' 0.1 ' ' 1
##
## Residual standard error: 12.63 on 25 degrees of freedom
## Multiple R-squared:  0.5696, Adjusted R-squared:  0.5352
## F-statistic: 16.54 on 2 and 25 DF,  p-value: 2.65e-05
```

(d) From the summary coefficients, we have:

$$\hat{Y} = 55.8495 + 2.9089 X_1 + 21.4332 X_2.$$

(e) We first note that the intercept coefficient is significant, indicating that the interpretation is valid. The expected salary is predicted to be \$55,850 when all predictor variables are 0. Note that the dummy variable (X_2) is 0 when an employee has an undergraduate degree.

(f) Since both variable coefficient p-values indicate statistical significance, it is appropriate to interpret the variable coefficients.

- For every additional year of experience, the annual salary is expected to increase by \$2909 assuming the degree variable remains constant.
- Since 1 designates that an employee has a graduate degree, the coefficient indicates an increase of \$21,433 in annual salary if someone has a graduate degree assuming the experience variable remains constant.

(g) The model fit proves reasonably well given that the R^2 is 0.5696, indicating the regression model explains approximately 57% of the variation in the annual salary.

7.4 Employee Salary Application: Dummy Variables 2

In this application, we will demonstrate how to modify the model interpretation by changing the dummy variable. Use the data from Table 7.1 to do the following.

(a) Create a new dummy variable that is 1 if someone has an undergraduate degree, and 0 if someone has a graduate degree.
(b) Fit a linear regression using this new dummy variable and experience to predict salary.
(c) Interpret the intercept coefficient.
(d) Interpret the variable coefficients.
(e) Discuss the differences between this model and the first application model.

Solution

(a) The dummy variable can easily be coded manually but also by using the `ifelse` command. The results are printed for the reader to verify the correct representation.

```
degree_dummy2 = ifelse(degree_dummy == 1, 0, 1)
degree_dummy2
```

```
## [1] 0 0 0 0 1 1 0 0 0 0 0 1 1 1 0 0 0 0 1 1 0 0 00 0 1 1 0
```

(b) The regression is fit and summarized using the commands below:

```
reg2 = lm(salary ~ experience + degree_dummy2)
summary(reg2)
```

```
##
## Call:
## lm(formula = salary ~ experience + degree_dummy2)
##
## Residuals:
##     Min      1Q  Median      3Q     Max
## -28.554  -6.356   0.082   4.264  32.173
##
## Coefficients:
##                   Estimate Std. Error t value Pr(>|t|)
## (Intercept)        77.2828     3.1086  24.861  < 2e-16 ***
## experience          2.9089     0.5771   5.040 3.36e-05 ***
## degree_dummy2     -21.4332     5.3113  -4.035 0.000452 ***
## ---
## Signif. codes:
## 0 '***' 0.001 '**' 0.01 '*' 0.05 '.' 0.1 ' ' 1
##
## Residual standard error: 12.63 on 25 degrees of freedom
## Multiple R-squared:  0.5696, Adjusted R-squared:  0.5352
## F-statistic: 16.54 on 2 and 25 DF,  p-value: 2.65e-05
```

(c) As in the previous problem, all three coefficient p-values indicate statistical significance; hence, it is appropriate to interpret the intercept and both variable coefficients. The expected salary is predicted to be $77,283 when all predictor variables are 0. Note that the dummy variable (X_2) is 0 when an employee has a graduate degree.

(d) The variable coefficient interpretations are:

- The annual salary coefficient interpretation is the same as in the previous application.
- Since one designates that an employee has an undergraduate degree, the coefficient indicates a decrease of $21,433 in annual salary if someone has an undergraduate degree, assuming the experience variable remains constant.

(e) The model in this application proves a similar overall fit to the previous application model as shown by the R_a^2 and R^2. The sign of the dummy coefficient is now the negative of the previous dummy coefficient. In addition, the intercept has changed to reflect the base level of employees with a graduate degree. Despite these differences, both models show the expected salary for an employee with no experience is $55,850 for those with an undergraduate degree and $77,283 for those with a graduate degree.

7.5 Multilevel Categorical Variables

If a categorical variable has k levels, or categories, $k - 1$ dummy variables are required to adequately reflect the information in the categorical variable. For example, a variable with levels A, B, and C could be represented by the dummy variables x_1 and x_2. If x_1 and x_2 are assigned as

$$x_1 = \begin{cases} 1 \text{ if level B} \\ 0 \text{ otherwise} \end{cases}$$

$$x_2 = \begin{cases} 1 \text{ if level C} \\ 0 \text{ otherwise} \end{cases}$$

then (x_1, x_2) will have values of $(0, 0)$ when level A occurs, $(1, 0)$ for when level B occurs, and $(0, 1)$ when level C occurs. Notice that since there were three levels, it was necessary to have two dummy variables to account for all three levels. Similarly, if a categorical variable has 10 levels, then 9 dummy variables would be necessary to extract all the information from the categorical variable.

The following application demonstrates how to code dummy variables with more than two categories.

7.6 Employee Salary Application: Dummy Variables with Multiple Levels

Continuing with the salary data (Table 7.1) from the previous two applications, use the data to do the following.

(a) Code the "Degree" variable as a vector of text strings.
(b) Fit a linear regression using degree and experience to predict salary.
(c) Explicitly state the linear regression equation.
(d) Interpret the intercept coefficient.
(e) Interpret the variable coefficients.
(f) Discuss the differences between this model and the previous application model.

Solution

(a) A text string can be denoted using double quotes or single quotes. Using the c
    function, a vector of text strings is defined by:

```
deg_string = c("MSBA", "MS", "MS", "MSBA", "BA", "BA", "MSBA",
               "MSBA", "MSBA", "MS", "MSBA", "BA", "BA", "BA",
               "MSBA", "MS", "MS", "MSBA", "BA", "BA", "MSBA",
               "MSBA", "MS", "MS", "MSBA", "BA", "BA", "MSBA")
```

(b) The lm function automatically converts the deg_string variable to dummy
    variables, making its usage similar to that of the previous application:

```
reg = lm(salary ~ experience + deg_string)
summary(reg)
```

```
##
## Call:
## lm(formula = salary ~ experience + deg_string)
##
## Residuals:
##     Min      1Q  Median      3Q     Max
## -19.989  -5.088  -2.453   5.091  29.116
##
## Coefficients:
##                  Estimate Std. Error t value Pr(>|t|)
## (Intercept)       55.0931     4.3599  12.636 4.26e-12 ***
## experience         3.0791     0.5129   6.004 3.38e-06 ***
## deg_stringMS      12.2638     5.6886   2.156   0.0413 *
## deg_stringMSBA    27.4550     5.1433   5.338 1.77e-05 ***
## ---
## Signif. codes:
## 0 '***' 0.001 '**' 0.01 '*' 0.05 '.' 0.1 ' ' 1
##
## Residual standard error: 11.15 on 24 degrees of freedom
## Multiple R-squared:  0.6782, Adjusted R-squared:  0.6379
## F-statistic: 16.86 on 3 and 24 DF,  p-value: 4.171e-06
```

(c) From the summary coefficients, we have:

$$\hat{Y} = 55.0931 + 3.0791\,X_1 + 12.2638\,X_2 + 27.4550\,X_3,$$

where

- X_1 is the years of experience.

- X_2 is a dummy denoting if an employee has a M.S. degree.
- X_3 is a dummy denoting if an employee has a M.S.B.A. degree.

(d) The p-value of the intercept indicates statistical significance, which means the intercept can be interpreted. The interpretation should state the following.

An employee with no experience who has a B.A. degree is expected to earn $55,093.

(e) All variable coefficient p-values indicate statistical significance; hence, it is appropriate to interpret all three variable coefficients.

- For every additional year of experience, the annual salary is expected to increase by $3079 assuming all other variables remain constant.
- If someone has a M.S. degree, then their salary is expected to be $12,264 higher than someone with a B.A. degree assuming all other variables remain constant.
- If someone has a M.S.B.A. degree, then their salary is expected to be $27,455 higher than someone with a B.A. degree assuming all other variables remain constant.

(f) Since the number of predictor variables increased in this application, the R^2 value will certainly increase. Therefore, we should look at R_a^2 to compare models rather than R^2. The model in the previous application had R_a^2 of 0.5352; the current model's R_a^2 of 0.6379 hence proves a better fit.

7.7   Coding Dummy Variables

In the first application, the dummy variable for noting if an employee had a graduate degree (degree_dummy) was created by replacing "Graduate" with 1 and "Undergraduate" with 0. In the third application, we took advantage of the lm function's ability to automatically transform the vector of text strings (deg_string) to dummy variables. When R automatically created these dummy variables, a two-step process took place. First, deg_string was converted into a factor variable. Second, the factor variable is then converted into dummy variables. In many instances, manually coding a variable as a factor variable proves advantageous. This can be done in R using the factor function.

```
deg_factor = factor(deg_string)
```

The deg_string vector above occurs as a character string. A convenient method for finding the variable type in R is to use the str function.

```
str(deg_string)
```

```
##   chr [1:28] "MSBA" "MS" "MS" "MSBA" "BA" "BA" "MSBA" ...
```

The "chr [1:28]" output denotes that the deg_string is a character string with 28 vector elements. The remaining output refers to the contents of the vector. The typeof function can also be used here, but the structure (str) function provides more information and can be used to inspect the variables within a data frame. Below, a dataframe is inspected using the structure function and using the data.frame command to create a dataframe with the variables experience, salary, and deg_string.

```
df = data.frame(experience, salary, deg_string)
str(df)
```

```
## 'data.frame':    28 obs. of  3 variables:
##  \$ experience: num  2 3 8 0 2 0 0 0 5 0 ...
##  \$ salary    : num  85 74 100 78 52 60 85 74 124 73 ...
##  \$ deg_string: chr  "MSBA" "MS" "MS" "MSBA" ...
```

From this output, we observe that the dataframe created has 28 observations with 3 variables. Both experience and salary are shown to contain numeric values as indicated by the "num" type. The deg_string vector is shown to contain text strings as denoted by the "chr" type.

The variable type of deg_string stands in contrast to the variable type of deg_factor:

```
str(deg_factor)
```

```
## Factor w/ 3 levels "BA","MS","MSBA": 3 2 2 3 1 1 3 3 3 2 ...
```

This output denotes a factor variable with 3 levels corresponding to a B.A. degree, a M.S. degree, and a M.S.B.A. degree. Each level corresponds to a different integer. Since B.A. is mentioned first, the integer 1 is associated with B.A. while M.S. and M.S.B.A. correspond to integers 2 and 3, respectively. This arrangement can be confirmed by observing the ordered integers of the factor variable: 3 2 2 3 1 ... align with the entries from deg_string: "MSBA" "MS" "MS" "MSBA" "BA"....

By default, the levels are decided by ordering them alphabetically and then assigning integers starting with 1. This step results in the first level being the reference or base level. When fitting a regression model with dummy variables, each dummy variable occurs in reference to another level. As mentioned in the previous application, the dummy variable for M.S.B.A. is interpreted in reference to someone with a B.A. degree. One helpful technique in dealing with factor variables consists of the ability to relevel the factors:

```
deg_factor2 <- relevel(deg_factor, ref = 3)
```

In the previous code, we specify the reference level to be the previous third level (M.S.B.A.). The structure of the new variable becomes:

```
str(deg_factor2)
```

```
## Factor w/ 3 levels "MSBA","BA","MS": 1 3 3 1 2 2 1 1 1 3 ...
```

While this new variable appears to be different, careful inspection reveals the sequence of this variable is still: "MSBA" "MS" "MS" "MSBA" "BA"....

Releveling a factor variable will change the regression analysis coefficients and the interpretation of the coefficients, but will not change the significance of the model or the overall characteristics of the model. As an alternative to releveling, defining a factor variable with the correct level sequence can be done using the `levels` option in the `factor` function:

```
deg_factor3 = factor(deg_string,
                     levels = c("MSBA","BA","MS"))
```

The structure of `deg_factor2` is identical to that of `deg_factor3`.

In most applications, manually coding dummy variables from factor variables is not necessary. Factor variable coding is typically enough to model a data set with linear regression. However, coding dummy variables manually can be done easily in R and several functions and packages exist to automatically code dummy variables. Two notable packages for this purpose include `fastDummies` or `dummies`.

7.8 Employee Salary Application: Dummy Variable Coding

With the salary data set, do the following.

(a) Create a new degree factor variable with M.S. as the base level.
(b) Fit a linear regression using this new factor variable and experience to predict salary.
(c) Interpret the intercept coefficient.
(d) Interpret the variable coefficients.
(e) Discuss the differences between this model and the previous application model.

Solution

(a) Rather than using the `relevel` function, we specify the `levels` argument within the `factor` function with the first level being a M.S. degree. Recall that R uses the first level as the base level when using the `lm` function.

```
deg_factor3 = factor(deg_string,
                        levels = c("MS", "MSBA", "BA"))
str(deg_factor)
```

```
## Factor w/ 3 levels "BA","MS","MSBA": 3 2 2 3 1 1 3 3 3 2 ...
```

(b) The model is fit with the new factor variable as follows:

```
reg = lm(salary ~ experience + deg_factor3)
summary(reg)
```

```
##
## Call:
## lm(formula = salary ~ experience + deg_factor3)
##
## Residuals:
##     Min      1Q  Median      3Q     Max
## -19.989  -5.088  -2.453   5.091  29.116
##
## Coefficients:
##                   Estimate Std. Error t value Pr(>|t|)
## (Intercept)        67.3569     4.4382  15.177 8.41e-14 ***
## experience          3.0791     0.5129   6.004 3.38e-06 ***
## deg_factor3MSBA    15.1912     5.3392   2.845  0.00894 **
## deg_factor3BA     -12.2638     5.6886  -2.156  0.04134 *
## ---
## Signif. codes:
## 0 '***' 0.001 '**' 0.01 '*' 0.05 '.' 0.1 ' ' 1
##
## Residual standard error: 11.15 on 24 degrees of freedom
## Multiple R-squared:  0.6782, Adjusted R-squared:  0.6379
## F-statistic: 16.86 on 3 and 24 DF,  p-value: 4.171e-06
```

(c) The p-value of the intercept indicates statistical significance, which means that the intercept can be interpreted. The interpretation holds that an employee with no experience who has a M.S. degree is expected to earn $67,357.
(d) All variable coefficient p-values indicate statistical significance so it is appropriate to interpret all three variable coefficients.

- The interpretation of the experience coefficient remains unchanged from that of the previous application.
- If someone has a M.S.B.A. degree, then their salary is expected to be $15,191 higher than someone with a M.S. degree assuming all other variables remain constant.

- If someone has a B.A. degree, then their salary is expected to be $12,264 lower (or −$12,264 higher) than someone with a M.S. degree, assuming all other variables remain constant.

(e) The model in this application proves a similar overall fit to the previous application model as shown by the R_a^2 and R^2. The dummy coefficients and the intercept have changed to reflect the base level of employees with a M.S. degree. Despite these differences, both models show that the expected salary for an employee with no experience is $55,093 for those with a B.A. degree, $67,357 for those with a M.S. degree, and $82,548 for those with a M.S.B.A. degree.

7.9 Modeling Curvilinear Relationships

When using linear regression, one makes the assumption that the data follows a linear relationship, but this scenario may not always be the case. Linear regression, despite the name, can handle nonlinear data by transforming the nonlinear components of the data.

Consider the data given in Fig. 7.1, which shows a nonlinear relationship between X and Y. The relationship in each scatterplot is said to be curvilinear. When we use a simple linear regression model to fit the data, we do not properly capture the pattern as shown by the blue line. Therefore, it becomes necessary to fit the data to a quadratic model by introducing an X^2 term. This relationship is shown in Fig. 7.1 by the red curve. The quadratic model can be written as:

$$Y = \beta_0 + \beta_1 X + \beta_2 X^2 + \varepsilon, \tag{7.1}$$

where β_0, β_1, and β_2, are the coefficient parameters of the model and ε is the error term. Fitting this equation to a data set provides the regression equation:

$$\hat{Y} = b_0 + b_1 X + b_2 X^2, \tag{7.2}$$

where b_0, b_1, and b_2 are the estimated coefficients.

7.10 Sales Performance Application: Quadratic Modeling

A manufacturer of quantum computing servers wants to investigate the relationship between the length of employment of their salespeople and the number of servers sold. The data in Table 7.2 lists the number of months each salesperson has been employed by the firm, "Employment," and the number of servers sold, "Sales," by 21 randomly selected salespeople.

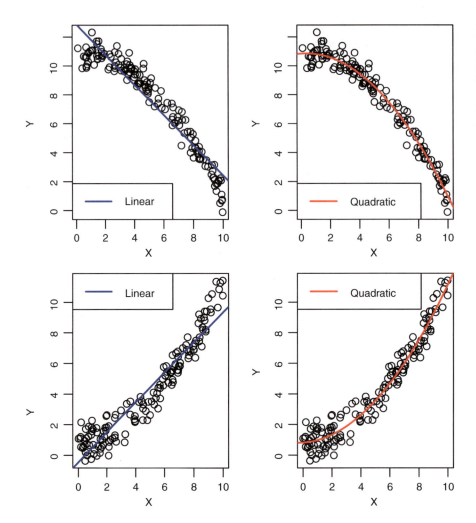

Fig. 7.1 Scatterplots of curvilinear data

If the response variable is Sales and the predictor variable is Employment, then do the following.

(a) Fit a simple linear regression line to the data and print the model summary.
(b) Plot the data in a scatterplot superimposing the simple linear regression line on the plot.
(c) Fit a quadratic regression model to the data and produce the corresponding summary.
(d) Plot the data in a scatterplot superimposing the quadratic model curve on the plot.
(e) Explain which relationship (simple linear or quadratic) more accurately reflects the relationship between "Sales" and "Employment."

Table 7.2 Sales data

Sales	Employment
379	57
239	114
225	100
240	58
403	75
227	46
338	106
5	2
167	50
284	77
378	93
399	70
239	106
306	89
325	86
113	33
270	67
131	24
30	5
125	15
267	52

Solution

(a) The summary can be found using the code below:

```
df = read.csv("Employment.csv")
reg = lm(Sales ~ Employment, data = df)
summary(reg)
```

```
##
## Call:
## lm(formula = Sales ~ Employment, data = df)
##
## Residuals:
##      Min       1Q   Median       3Q      Max
## -127.812  -55.816    0.298   26.631  151.518
##
## Coefficients:
##              Estimate Std. Error t value Pr(>|t|)
## (Intercept)   88.1515    38.4925   2.290 0.033628 *
## Employment     2.4444     0.5412   4.517 0.000236 ***
## ---
```

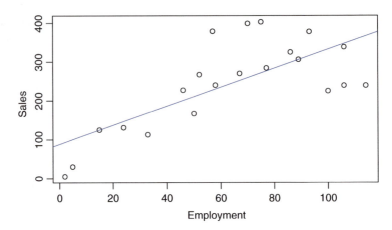

Fig. 7.2 Sales and employment: linear trend

```
## Signif. codes:
## 0 '***' 0.001 '**' 0.01 '*' 0.05 '.' 0.1 ' ' 1
##
## Residual standard error: 81.41 on 19 degrees of freedom
## Multiple R-squared:  0.5178, Adjusted R-squared:  0.4924
## F-statistic:  20.4 on 1 and 19 DF,  p-value: 0.0002361
```

(b) The `abline` function from Chap. 3 allows us to superimpose the line on the
plot. Using the `col` option within the `abline` function allows us to color the
line blue (Fig. 7.2).

```
plot(df$Employment, df$Sales, xlab = "Employment",
     ylab = "Sales")
abline(reg, col = "blue")
```

(c) The `I` function can be used within a formula to include a squared term in the
model as in the following:

```
quad = lm(Sales ~ Employment + I(Employment^2), data = df)
summary(quad)
```

```
##
## Call:
## lm(formula = Sales ~ Employment + I(Employment^2),
## data = df)
##
## Residuals:
##     Min      1Q  Median      3Q     Max
## -96.041 -33.745  -6.939  37.994  95.651
```

```
##
## Coefficients:
##                   Estimate Std. Error t value Pr(>|t|)
## (Intercept)      -26.44537   39.28955  -0.673 0.509441
## Employment         8.32360    1.46025   5.700 2.09e-05 ***
## I(Employment^2)   -0.05068    0.01212  -4.183 0.000559 ***
## ---
## Signif. codes:
## 0 '***' 0.001 '**' 0.01 '*' 0.05 '.' 0.1 ' ' 1
##
## Residual standard error: 59.56 on 18 degrees of freedom
## Multiple R-squared:  0.7555, Adjusted R-squared:  0.7283
## F-statistic:  27.8 on 2 and 18 DF,  p-value: 3.127e-06
```

(d) To plot the quadratic curve, we employ the `lines` function, which connects the points with line segments. First, we create a vector of X values (`xvalues`) starting at the minimum value of 2, incrementing by 1 until we reach the maximum value of 114. Then, we place this vector in a dataframe with the column label of "Employment" since that is the name of the predictor variable. The `predict` function requires a model object and a dataframe from which to get predictions. The `lines` function requires we input a vector of X values followed by a vector of Y values. For aesthetics, the quadratic curve is colored red using the `col` option (Fig. 7.3).

```
plot(df$Employment, df$Sales, xlab = "Employment",
     ylab = "Sales")
xvalues = 2:114
xvalues_df = data.frame(Employment = xvalues)
```

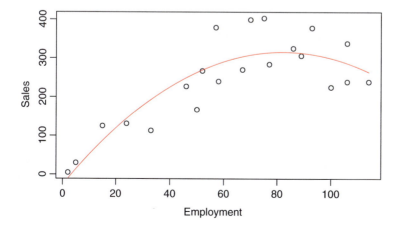

Fig. 7.3 Sales and employment: curvilinear trend

```
pred = predict(quad, xvalues_df)
lines(xvalues, pred, col = "red")
```

(e) The superimposed model fit line and the curve in the scatterplots both indicate strong fits. However, the relationship between the length of time employed and the number of sales is better captured using the quadratic model. The R^2_a values in the summaries also show the quadratic model proves a better fit since 0.7283 is greater than 0.4924.

7.11 Mean-Centering

In many cases, the intercept does not reveal anything interesting about the data, or worse, the intercept does not make sense. These scenarios can be attributed to the intercept reflecting the expected value of the response when all of the predictor variables are zero. If a predictor variable being zero does not make sense, then the corresponding prediction will not make sense either.

Mean-centering a predictor variable allows the variable to retain the spread, but the value of the mean-centered variable occurs in reference to the mean. To mean-center the variable X, simply subtract the mean from each observation. If x_i represents the ith value of X, then the ith transformed value is:

$$x_i - \bar{x}. \tag{7.3}$$

Note the mean of the transformed data will be zero, which allows for a convenient interpretation of the intercept. The application below provides an example of how mean-centering can be utilized.

7.12 Marketing Toys Application: Mean-Centering

A popular toy company sells two different toys: toy A and toy B. The company wishes to know how much to spend on marketing to achieve their sales goals. The data are in the file: MarketingToys.csv. Units of Sales and Marketing are in thousands of USD. The data can be loaded into R using read.csv. Additionally, we observe the first 6 observations:

```
df = read.csv("MarketingToys.csv")
head(df)
```

```
##   Toy   Sales Marketing
## 1   A 855.844   174.081
```

```
## 2    A 728.745    175.027
## 3    B 887.742    180.757
## 4    B 604.876    168.615
## 5    B 703.407    162.877
## 6    A 579.380    164.438
```

After loading in the data, complete the following.

(a) Create scatterplots of the marketing expenditures versus the sales revenue for all observations, toy A only, and toy B only.
(b) Fit a model and print a summary predicting sales as a function of marketing and the toy type.
(c) Interpret the coefficients.
(d) Mean center the marketing variable and repeat part b.
(e) Interpret the intercept coefficient for the model in part d.
(f) Change the base level, produce a summary and state the new meaning of the intercept.

Solution

(a) The overall scatterplot is given using the `plot` command. The `xlab` and `ylab` options are specified to produce a well-labeled plot (Fig. 7.4):

```
plot(df$Marketing, df$Sales, xlab = "Marketing (USD)",
     ylab = "Sales (USD)")
```

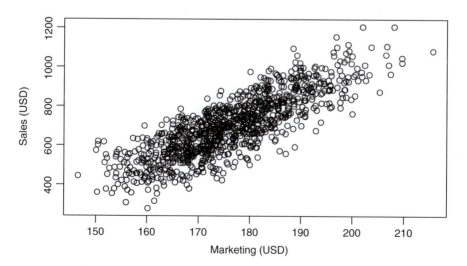

Fig. 7.4 Marketing and sales of products A and B

A few coding details regarding the next code chunk:

(1) For coding ease, the dataframe df is broken up into two data frames using simple indexing. The resulting df_A and df_B represent dataframes of products A and B, respectively.
(2) Here, we introduce par to allow for two graphs to be posted side by side. The mfrow = c(1,2) option within par specifies 1 row and 2 columns of graphs.
(3) Within the plot command, we specify the limits of the X and Y axes using xlim and ylim. Setting the upper and lower limits of each axis allows for a direct visual comparison between both products.
(4) Toys A and B are colored as "navy blue" and "maroon," respectively (Fig. 7.5).

```
df_A = df[df$Toy == "A",]
df_B = df[df$Toy == "B",]

par(mfrow=c(1,2))
plot(df_A$Marketing, df_A$Sales, xlab = "Marketing",
ylab = "Sales", xlim = c(120, 220),
ylim = c(300, 2200), col = "navyblue")

plot(df_B$Marketing, df_B$Sales, xlab = "Marketing",
ylab = "Sales", xlim = c(120, 220),
ylim = c(300, 2200), col = "maroon")
```

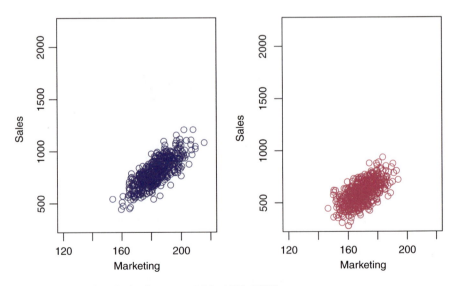

Fig. 7.5 Sales and marketing by toy model (in 1000s USD)

(b) Using the dataframe df, a fit can easily be made using the lm function:

```
reg1 = lm(Sales ~ Marketing + Toy, data = df)
summary(reg1)
```

```
##
## Call:
## lm(formula = Sales ~ Marketing + Toy, data = df)
##
## Residuals:
##     Min     1Q  Median     3Q     Max
## -253.04  -55.00   -3.60   60.37  245.84
##
## Coefficients:
##                Estimate Std. Error t value Pr(>|t|)
## (Intercept)  -907.5484    52.7491  -17.20   <2e-16 ***
## Marketing       9.3253     0.2875   32.43   <2e-16 ***
## ToyB          -76.8916     6.5574  -11.73   <2e-16 ***
## ---
## Signif. codes:
## 0 '***' 0.001 '**' 0.01 '*' 0.05 '.' 0.1 ' ' 1
##
## Residual standard error: 83.74 on 997 degrees of freedom
## Multiple R-squared:  0.7158, Adjusted R-squared:  0.7152
## F-statistic:  1255 on 2 and 997 DF,  p-value: < 2.2e-16
```

(c) The interpretation of the coefficients is as follows:

- Toy B has on average $76,892 less sales than toy A, or alternatively, toy A has sales on average $76,892 more than toy B.
- For every $1000 increase in marketing expenditures, there is expected to be $9325 more sales.
- The intercept is the value when all variables are 0 (0 indicates toy A the ToyB dummy). Note that it would not make sense for sales to be negative which implies the intercept should not be interpreted.

(d) Subtract the mean of the marketing expenditures from all the marketing values. The summary displays this new variable and confirms the new mean is zero:

```
df$Mktg_MC = df$Marketing - mean(df$Marketing)
summary(df$Mktg_MC)
```

```
##     Min.  1st Qu.   Median     Mean  3rd Qu.     Max.
## -29.8013  -8.2923  -0.8773   0.0000   7.3855  39.3897
```

The regression is now fit using the mean-centered variable with Toy:

```
reg2 = lm(Sales ~ Mktg_MC + Toy, data = df)
summary(reg2)
```

```
##
## Call:
## lm(formula = Sales ~ Mktg_MC + Toy, data = df)
##
## Residuals:
##      Min      1Q  Median      3Q      Max
## -253.04  -55.00   -3.60   60.37   245.84
##
## Coefficients:
##               Estimate Std. Error t value Pr(>|t|)
## (Intercept) 736.2614      4.2145  174.70   <2e-16 ***
## Mktg_MC       9.3253      0.2875   32.43   <2e-16 ***
## ToyB        -76.8916      6.5574  -11.73   <2e-16 ***
## ---
## Signif. codes:
## 0 '***' 0.001 '**' 0.01 '*' 0.05 '.' 0.1 ' ' 1
##
## Residual standard error: 83.74 on 997 degrees of freedom
## Multiple R-squared:  0.7158, Adjusted R-squared:  0.7152
## F-statistic:  1255 on 2 and 997 DF,  p-value: < 2.2e-16
```

(e) In the previous part, a zero for both variables indicates that the sales from toy A without any marketing expenses will be $736,261.
(f) To change the base level, we redefine the Toy variable with B as the base level and then fit the regression model:

```
df$Toy = factor(df$Toy, levels = c("B","A"))
reg3 = lm(Sales ~ Mktg_MC + Toy, data = df)
summary(reg3)
```

```
##
## Call:
## lm(formula = Sales ~ Mktg_MC + Toy, data = df)
##
## Residuals:
##      Min      1Q  Median      3Q      Max
## -253.04  -55.00   -3.60   60.37   245.84
##
## Coefficients:
```

```
##              Estimate Std. Error t value Pr(>|t|)
## (Intercept) 659.3698     4.2145  156.45  <2e-16 ***
## Mktg_MC        9.3253     0.2875   32.43  <2e-16 ***
## ToyA          76.8916     6.5574   11.73  <2e-16 ***
## ---
## Signif. codes:
## 0 ’***’ 0.001 ’**’ 0.01 ’*’ 0.05 ’.’ 0.1 ’ ’ 1
##
## Residual standard error: 83.74 on 997 degrees of freedom
## Multiple R-squared:  0.7158, Adjusted R-squared:  0.7152
## F-statistic:  1255 on 2 and 997 DF,  p-value: < 2.2e-16
```

In this case, a 0 for both variables indicates that the sales with mean marketing expenditures will be $659,370.

7.13 General Linear Regression Model

Models in which the parameters $(\beta_0, \beta_1, \ldots, \beta_p)$ all correspond to variables with exponents of one are called linear models. However, as shown in the quadratic model, it is possible to square the predictor variable and use the resulting square as another predictor variable. This method allows us to model nonlinear functions using linear regression techniques. Since the aforementioned transformed models can be represented as linear models, they are a class of regression models that are referred to as general linear models.

The general linear model involving p predictor variables is

$$Y = \beta_0 + \beta_1 Z_1 + \beta_2 Z_2 + \ldots + \beta_p Z_p + \varepsilon. \tag{7.4}$$

Each of the predictor variables Z_i is a function of $X_1, X_2, \ldots, X_k$ (the variables for which data have been collected).

The simplest case occurs when we have collected data for just one variable X_1 and want to estimate Y by using a straight-line relationship. In this case $Z_1 = X_1$. The resulting model, the simple linear regression model is:

$$Y = \beta_0 + \beta_1 X_1 + \varepsilon. \tag{7.5}$$

Within the framework of the general linear regression model, this model is called a simple first-order model with one predictor variable.

To account for a curvilinear relationship, we might set $Z_1 = X_1$ and $Z_2 = X_1^2$. The resulting quadratic model is

$$Y = \beta_0 + \beta_1 X_1 + \beta_2 X_1^2 + \varepsilon \tag{7.6}$$

and is called a simple second-order model with one predictor variable.

Table 7.3 General linear regression model versus generalized linear model

	General linear regression model	Generalized linear model
Typical estimation method	Least squares	Maximum likelihood or Bayesian
Examples	ANOVA, linear regression	Logistic regression, Poisson regression, gamma regression, general linear regression model
R function	lm()	glm()

If the original data set consists of observations for Y and two predictor variables X_1 and X_2, we might develop a second-order model with two predictor variables:

$$Y = \beta_0 + \beta_1 X_1 + \beta_2 X_2 + \beta_3 X_1^2 + \beta_4 X_2^2 + \beta_5 X_1 X_2 + \varepsilon. \tag{7.7}$$

In this model, we include the variable $Z_5 = X_1 X_2$ to account for the potential effects of the two variables acting together. This type of effect is called an interaction.

Sometimes, the general linear regression model is referred to as the general linear model, which should not be confused with the Generalized Linear Model (GLM). The GLM serves as a flexible generalization of linear regression that allows for response variables that have non-normal distributions. The GLM generalizes linear regression by using a link function on the response variable. These differences are summarized in Table 7.3.

7.14   Interactions

To simplify the discussion, we focus here on the case of two predictor variables with an interaction. The resulting model is therefore:

$$Y = \beta_0 + \beta_1 X_1 + \beta_2 X_2 + \beta_3 X_1 X_2 + \varepsilon. \tag{7.8}$$

If X_2 is a dummy variable, then the model changes for each outcome of X_2. When $X_2 = 0$, the model is:

$$Y = \beta_0 + \beta_1 X_1 + \varepsilon, \tag{7.9}$$

but when $X_2 = 1$, the model is:

$$Y = \beta_0 + \beta_1 X_1 + \beta_2 (1) + \beta_3 X_1 (1) + \varepsilon.$$

$$Y = (\beta_0 + \beta_2) + (\beta_1 + \beta_3) X_1 + \varepsilon. \tag{7.10}$$

Notice the intercept for Eq. (7.10) is $\beta_0 + \beta_2$ and the slope is $\beta_1 + \beta_3$. By including an interaction with a dummy variable, the interaction model is equivalent to creating two separate models for each category the dummy variable represents.

7.15 Marketing Toys Application: Interactions

Using the marketing toys data from the previous application, perform the following steps.

(a) Fit a model with the mean-centered marketing expenditures, the toy type, and an interaction as predictors of the sales. Use "B" as the base level for the dummy variable and print a summary of the fitted model.
(b) Interpret the coefficient of the interaction.
(c) Use the summary to find the fitted models that would result if we were to fit models for type A and type B separately.
(d) Plot the mean-centered marketing variable on the x-axis and the sales on the y-axis, then plot lines representing both toy types.
(e) Fit a model using only the type B to predict sales using marketing. How do the coefficient values differ from those in the model with the interaction?

Solution

(a) Using the mean-centered marketing variable and the interaction Mktg_MC:Toy, the model is fit:

```
reg = lm(Sales ~ Mktg_MC + Toy + Mktg_MC:Toy, data = df)
summary(reg)
```

```
##
## Call:
## lm(formula = Sales ~ Mktg_MC + Toy + Mktg_MC:Toy, data =
## df)
##
## Residuals:
##       Min       1Q    Median       3Q      Max
## -255.851  -54.512    -2.249   61.757  242.921
##
## Coefficients:
##                 Estimate Std. Error t value Pr(>|t|)
## (Intercept)     650.3707     4.7433 137.113  < 2e-16 ***
## Mktg_MC           7.9869     0.4383  18.223  < 2e-16 ***
## ToyA             79.2695     6.5348  12.130  < 2e-16 ***
## Mktg_MC:ToyA      2.3232     0.5774   4.023 6.18e-05 ***
## ---
```

```
## Signif. codes:
## 0 '***' 0.001 '**' 0.01 '*' 0.05 '.' 0.1 ' ' 1
##
## Residual standard error: 83.11 on 996 degrees of freedom
## Multiple R-squared:  0.7203, Adjusted R-squared:  0.7195
## F-statistic: 855.1 on 3 and 996 DF,  p-value: < 2.2e-16
```

(b) The interaction can be interpreted as:

For every dollar spent on marketing, there is an additional $2.32 increase in sales for toy A as compared to toy B.

(c) From the summary, when ToyA is 0, the dummy variable and interaction coefficient can be disregarded. If X and Y denote mean-centered marketing and sales in thousands of USD, the resulting model for toy B is:

$$\hat{Y} = 650.3707 + 7.9869X.$$

The intercept for the toy A model can be found by adding the coefficient of the ToyA dummy variable to the intercept. The interaction term of the model can be added to the slope of the toy B model to get the slope of the toy A model. If X and Y denote mean-centered marketing and sales, respectively, the resulting model for toy A is:

$$\hat{Y} = (650.3707 + 79.2695) + (7.9869 + 2.3232)X$$

$$\hat{Y} = 729.6402 + 10.3101X.$$

(d) Altering the plot code in the previous application to contain the mean-centered marketing expenditures on the x-axis, we have:

```
plot(df$Mktg_MC, df$Sales, xlab = "Mean-centered Marketing",
     ylab = "Sales")
abline(a = 650.3707, b = 7.9869, col = "maroon", lwd = 2)
abline(a = 729.6402, b = 10.3101, col = "navyblue", lwd = 2)
legend('topleft', c('Toy A','Toy B'), lty = c(1,1),
       col = c("navyblue", "maroon"), lwd = 2)
```

In the above code, we explicitly state the options of the abline function to match the intercept (a) and the slope (b) for each model. We use the lwd option to increase the width of the line. We also add a legend to clarify which line denotes each toy. The result is shown in Fig. 7.6.

(e) Fitting a model for Toy A using the mean-centered marketing variable is done using the following code.

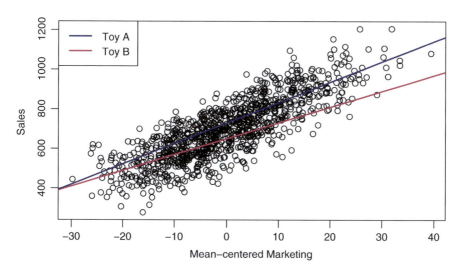

Fig. 7.6 Scatterplot of toys

```
df_A$Mktg_MC = df_A$Marketing - mean(df_A$Marketing)
reg = lm(Sales ~ Mktg_MC, data = df_A)
summary(reg)
```

```
##
## Call:
## lm(formula = Sales ~ Mktg_MC, data = df_A)
##
## Residuals:
##      Min       1Q    Median       3Q      Max
## -255.851  -53.418   -2.728   58.221  240.489
##
## Coefficients:
##              Estimate Std. Error t value Pr(>|t|)
## (Intercept) 798.9623     3.6056  221.59   <2e-16 ***
## Mktg_MC      10.3100     0.3647   28.27   <2e-16 ***
## ---
## Signif. codes:
## 0 '***' 0.001 '**' 0.01 '*' 0.05 '.' 0.1 ' ' 1
##
## Residual standard error: 80.62 on 498 degrees of freedom
## Multiple R-squared:  0.6161,  Adjusted R-squared:  0.6153
## F-statistic: 799.1 on 1 and 498 DF,  p-value: < 2.2e-16
```

For the toy A model, the slope is the same as the result from part c. Since marketing was centered based upon the marketing for Toy A alone, the intercept differs between the model given here and from part c.

7.16 Case Study: Social Media

7.16.1 Problem Statement

Social media has become a driving force in the world of business in recent years. It provides users with a forum to express their views, as well as to teach and learn from other users. Yet some posts do not get much notice while others go viral. Being able to create popular posts can be a very profitable skill, as seen by the revenue generated by many influencers who are able to regularly produce attention-grabbing posts.

Luke is the founder and CEO of a small video marketing and production company. He currently manages many of his clients' videos and distributes them on a popular social media platform. Luke measures the success of his videos by the number of "likes" that the videos get, since the number of likes has a direct relationship with the profit from ads that will be collected. After collecting some data, Luke seeks to find which promoter he should use, and the effects the sentiment and the age of a video have on the number of likes.

Here we will use regression analysis to help Luke understand his data and, therefore, generate larger profits for his clients.

7.16.2 Data Description

Luke was able to collect a small data set consisting of 157 videos containing 4 variables. The variables are listed below.

- Likes—The number of likes a video received.
- Promoter—A categorical variable representing the promoter corresponding to the video.
- Age—The age of the video in days from the reference date of August 10, 2021.
- Sentiment—The average sentiment score of the comments from the video. The sentiment score is calculated based on a sentiment scoring algorithm using a dictionary which maps positive or negative sentiments with each word. Negative values correspond to a negative sentiment, whereas positive values correspond to a positive sentiment. For more information on sentiment scoring algorithms, access the R help file for the get_sentiment function.

The data is contained within the Youtube.csv file. Here we explore the data set by loading in and obtaining the head of the dataframe.

```
df = read.csv("Youtube.csv")
head(df)
```

```
##      Likes Promoter Age    Sentiment
## 1 240372          A 437  0.115398920
## 2 100243          A 234  0.005290118
## 3  19015          A   4  0.022849940
## 4 188370          A 364 -0.003857647
## 5 173882          A 358 -0.016700042
## 6  26730          A  94 -0.001736301
```

From the head function, the Promoter variable has the incorrect variable type
(chr). Therefore, we convert the variable to a factor variable.

```
df$Promoter = factor(df$Promoter)
```

Next, we visualize the data using the plot function. For a more interesting plot,
the colors vector is created using an ifelse function from the "Promoter" variable
designating promoter "A" as "red," and promoter "B" as "blue." Then, we pass
along colors to the col argument within the plot function. The dataframe df
uses referencing to exclude column 2 since that column is represented by the red
and blue colors (Fig. 7.7).

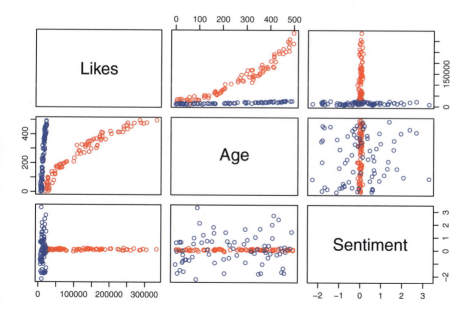

Fig. 7.7 Scatterplot matrix colored by promoters

```
colors = ifelse(df$Promoter == "A", "red", "blue")
plot(df[, -2], col = colors)
```

Investigating the scatterplot matrix above yields some interesting findings. Most notably, the Likes variable follows a different pattern for each promoter type since the red and blue observations are distinct.

7.16.3   Promoter A Model

Since the scatterplot matrix above has notably different characteristics for each promoter, we can segment the data by promoter which will allow us to analyze the data from each promoter separately.

First, we fit a multiple regression model on the dataframe for promoter "A" using all of the other variables to predict the Likes variable.

```
df_A = df[df$Promoter == "A",]
reg = lm(Likes ~ Age + Sentiment, data = df_A)
summary(reg)
```

```
##
## Call:
## lm(formula = Likes ~ Age + Sentiment, data = df_A)
##
## Residuals:
##     Min    1Q Median    3Q    Max
## -52926 -18793  -4955  18129  79420
##
## Coefficients:
##                 Estimate Std. Error t value Pr(>|t|)
## (Intercept) -20326.65    5665.94  -3.588 0.000566 ***
## Age             552.79      19.09  28.961  < 2e-16 ***
## Sentiment    -28295.58   54576.69  -0.518 0.605537
## ---
## Signif. codes:
## 0 '***' 0.001 '**' 0.01 '*' 0.05 '.' 0.1 ' ' 1
##
## Residual standard error: 25230 on 82 degrees of freedom
## Multiple R-squared:  0.9133, Adjusted R-squared:  0.9112
## F-statistic:   432 on 2 and 82 DF,  p-value: < 2.2e-16
```

This summary has a high R_a^2 with the Age being significant and Sentiment being insignificant according to the p-values. Recalling from the scatterplot matrix, the relationship between Likes and Age is curvilinear. Hence, we add another variable for the squared Age to the regression formula.

```
reg=lm(Likes~Age+I(Age^2)+Sentiment, data = df_A)
summary(reg)
```

```
##
## Call:
## lm(formula = Likes ~ Age + I(Age^2) + Sentiment,
## data = df_A)
##
## Residuals:
##     Min    1Q Median    3Q    Max
## -37536  -9954  -1165   8418  37159
##
## Coefficients:
##                Estimate Std. Error t value Pr(>|t|)
## (Intercept) 2.307e+04  5.045e+03   4.574  1.7e-05 ***
## Age         2.831e+01  4.592e+01   0.616    0.539
## I(Age^2)    1.045e+00  8.847e-02  11.808  < 2e-16 ***
## Sentiment   8.313e+02  3.338e+04   0.025    0.980
## ---
## Signif. codes:
## 0 '***' 0.001 '**' 0.01 '*' 0.05 '.' 0.1 ' ' 1
##
## Residual standard error: 15390 on 81 degrees of freedom
## Multiple R-squared:  0.9681, Adjusted R-squared:  0.967
## F-statistic: 820.7 on 3 and 81 DF,  p-value: < 2.2e-16
```

From the summary, the R_a^2 increased by more than 0.05; hence the model fit has improved. Next, we again analyze the p-values of the coefficients, with the largest p-value being that of the Sentiment variable. From this observation, we drop the Sentiment from the model and observe the resulting R_a^2.

```
reg=lm(Likes~I(Age^2)+Age, data = df_A)
summary(reg)
```

```
##
## Call:
## lm(formula = Likes ~ I(Age^2) + Age, data = df_A)
##
## Residuals:
##     Min    1Q Median    3Q    Max
## -37548 -10042  -1153   8462  37179
##
## Coefficients:
##                Estimate Std. Error t value Pr(>|t|)
```

```
## (Intercept) 2.306e+04  4.955e+03    4.653 1.24e-05 ***
## I(Age^2)     1.045e+00  8.769e-02   11.911  < 2e-16 ***
## Age          2.845e+01  4.531e+01    0.628    0.532
## ---
## Signif. codes:
## 0 '***' 0.001 '**' 0.01 '*' 0.05 '.' 0.1 ' ' 1
##
## Residual standard error: 15290 on 82 degrees of freedom
## Multiple R-squared:  0.9681, Adjusted R-squared:  0.9674
## F-statistic:  1246 on 2 and 82 DF,  p-value: < 2.2e-16
```

While removing Sentiment only marginally increases the R_a^2, the model complexity decreases which adds to its interpretability. Also, the p-value of the unsquared Age variable is insignificant according to its p-value; therefore, we remove this variable from the model as well.

```
reg = lm(Likes ~ I(Age^2), data = df_A)
summary(reg)
```

```
##
## Call:
## lm(formula = Likes ~ I(Age^2), data = df_A)
##
## Residuals:
##     Min     1Q Median     3Q     Max
## -37563  -9982   -874   8452   35600
##
## Coefficients:
##                   Estimate Std. Error t value Pr(>|t|)
## (Intercept) 2.571e+04  2.564e+03    10.03 5.72e-16 ***
## I(Age^2)     1.098e+00  2.191e-02    50.10  < 2e-16 ***
## ---
## Signif. codes:
## 0 '***' 0.001 '**' 0.01 '*' 0.05 '.' 0.1 ' ' 1
##
## Residual standard error: 15240 on 83 degrees of freedom
## Multiple R-squared:  0.968,  Adjusted R-squared:  0.9676
## F-statistic:  2510 on 1 and 83 DF,  p-value: < 2.2e-16
```

The R_a^2 again increases by a miniscule amount. However, by removing the unsquared variable from the model, the interpretation of the coefficient in the fitted model for promoter A is simplified significantly. The interpretation is:

For every unit increase in the squared age, the number of likes is expected to increase by 1.098.

The interpretation of the intercept for promoter A videos is:

When the video is first released (the age is zero days), the expected number of likes is 25,710.

7.16.4 Promoter B Model

A similar process can be done to get a model for promoter B. The resulting model is shown below in the summary here.

```
df_B = df[df$Promoter == "B",]
reg = lm(Likes ~ I(Age^2), data = df_B)
summary(reg)
```

```
##
## Call:
## lm(formula = Likes ~ I(Age^2), data = df_B)
##
## Residuals:
##     Min      1Q  Median      3Q     Max
## -3315.4  -901.5   109.4  1117.6  3942.9
##
## Coefficients:
##               Estimate Std. Error t value Pr(>|t|)
## (Intercept) 8.596e+03  2.886e+02   29.79   <2e-16 ***
## I(Age^2)    5.726e-02  2.599e-03   22.03   <2e-16 ***
## ---
## Signif. codes:
## 0 '***' 0.001 '**' 0.01 '*' 0.05 '.' 0.1 ' ' 1
##
## Residual standard error: 1696 on 70 degrees of freedom
## Multiple R-squared:  0.874,  Adjusted R-squared:  0.8722
## F-statistic: 485.5 on 1 and 70 DF,  p-value: < 2.2e-16
```

The interpretation of the Age-squared coefficient for the promoter B model is:

For every unit increase in the squared age, the number of likes is expected to increase by 0.05726.

The interpretation of the intercept for promoter B videos is:

When the video is first released (the age is zero days), the expected number of likes is 8596.

7.16.5 Combined Model

As shown earlier in this chapter, an interaction can be added to model the promoter A and B models jointly.

```
reg = lm(Likes ~ I(Age^2) + Promoter:I(Age^2) + Promoter,
         data = df)
summary(reg)
```

```
##
## Call:
## lm(formula = Likes ~ I(Age^2) + Promoter:I(Age^2) +
## Promoter, data = df)
##
## Residuals:
##    Min    1Q Median    3Q   Max
## -37563 -3310  -187  2800 35600
##
## Coefficients:
##                       Estimate Std. Error t value Pr(>|t|)
## (Intercept)          2.571e+04  1.898e+03  13.548  < 2e-16
## I(Age^2)             1.098e+00  1.622e-02  67.672  < 2e-16
## PromoterB           -1.712e+04  2.700e+03  -6.341 2.44e-09
## I(Age^2):PromoterB  -1.041e+00  2.371e-02 -43.892  < 2e-16
##
## (Intercept)         ***
## I(Age^2)            ***
## PromoterB           ***
## I(Age^2):PromoterB  ***
## ---
## Signif. codes:
## 0 '***' 0.001 '**' 0.01 '*' 0.05 '.' 0.1 ' ' 1
##
## Residual standard error: 11280 on 153 degrees of freedom
## Multiple R-squared:  0.982,  Adjusted R-squared:  0.9816
## F-statistic:  2781 on 3 and 153 DF,  p-value: < 2.2e-16
```

From this model with interactions, the R_a^2 has improved due to the number of model observations increasing. The coefficient for Age-squared remains 1.098. The interpretation of the coefficient for the interaction is:

For every unit increase in the squared age, there is a 1.041 decrease in Likes for promoter B videos as compared to promoter A videos.

Note that the intercept for promoter B videos is also less than that of the intercept for promoter A videos.

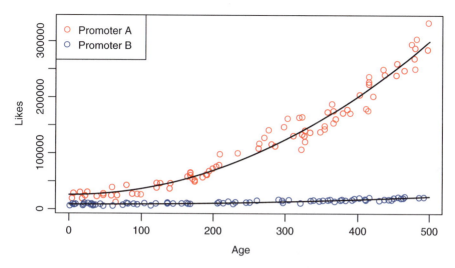

Fig. 7.8 Scatterplot colored by promoters

```
plot(Likes ~ Age, col = colors, data = df)

xvalues = 0:500
xvalues_df_A = data.frame(Age = xvalues, Promoter = "A")
pred = predict(reg, xvalues_df_A)
lines(xvalues, pred, lwd = 1.5)

xvalues = 0:500
xvalues_df_B = data.frame(Age = xvalues, Promoter = "B")
pred = predict(reg, xvalues_df_B)
lines(xvalues, pred, lwd = 1.5)

legend('topleft', c('Promoter A','Promoter B'), pch= 1,
       col = c("red", "blue"))
```

In Fig. 7.8, the promoter A and B videos are distinguished by red and blue colors, respectively. The Likes are plotted on the y-axis and the Age is plotted on the x-axis. The regression trends are plotted in black for both the promoter A model and the promoter B model.

7.16.6   Case Conclusion

From the results shown above, a clear distinction exists between the two promoters. Promoter A not only has the initial Likes but also has the stronger trend for the number of Likes when accounting for Age. Luke can confidently contract promoter A to advertise his clients' videos assuming the costs are not prohibitive. The quality of promotion can be continued, as promoter A has the capacity to handle all of the videos.

This case study demonstrates the utility of interactions and dummy variables within regression analysis. Observing the scatterplots carefully was integral, and allowed for the data and the models to be separated by promoter. This led us to create a model for each promoter separately and then a joint one to account for both promoters.

Problems

1. **Job Changes Data with Dummy Variables**
   Using the JobChanges.csv file, answer the following questions.

   a. Fit a simple linear regression model to predict the annual salary in thousands of USD (Salary) as a function of the number of job changes (Jobs). Print a summary of the model.
   b. What percent of the variation in Salary can be explained by the model in the previous part?
   c. Create a barplot of the 3 education levels in the Education variable by passing a summary of the Education factor into the barplot function.
   d. Fit a multiple regression model to predict the Salary using Jobs and 2 dummy variables representing education levels as predictor variables. Print a summary of the model.
   e. Create a scatterplot of Jobs versus Salary. Overlay the scatterplot with the regression line for the HS level as given in the previous model. Also, overlay the scatterplot with the regression lines for the other 2 levels as given by the previous model.
   f. How should the dummy variables be interpreted?

2. **Job Changes Data with Interactions**
   Using the JobChanges.csv file, answer the following questions.

   a. Fit a multiple regression model on the data set predicting Salary as a function of Jobs and Education. Include an interaction between the predictor variables. Print a summary of the model.
   b. Interpret the interaction coefficients from the previous model.
   c. Create a scatterplot of Jobs versus Salary. Overlay the scatterplot with the regression line for the HS level as given in the previous model. Also, overlay the scatterplot with the regression lines for the remaining levels as given by the previous model.
   d. Using the multiple regression model with the interaction from part a, specify the simple linear regression equation that predicts Salary using Jobs as a predictor for high school graduates (HS). Also, specify the same simple linear

regressions for college graduates (bachelors) and people with a graduate degree (masters).

e. Repeat the plot from part c and color the observations by Education level.

f. Create and add a variable called JobsM which mean-centers the Jobs variable.

g. Fit a multiple regression model which uses the following predictors:

- A dummy variable for observations at the bachelors level
- A dummy variable for observations at the masters level
- The mean-centered Jobs
- Interactions between each dummy variable and the mean-centered Jobs

h. Interpret the interactions from the previous part.

3. Automotive Tire Sales with Categorical Variables

Here we will use the tires data to solve the questions below. Load in the Tires.csv file by running the code chunk below.

```
## df = read.csv("TireSales.csv")
```

a. Get a summary of the data using the summary function.

b. List the variables that are numeric and the variables that are categorical.

c. Generate a linear model to predict the response variable (Sales) from all of the other variables in the data set.

d. Display a summary of the model given from the previous part.

e. Interpret the dummy variables from the summary in part d.

4. Facebook Analysis with Dummy Variables

Using the Facebook.csv file, answer the following questions.

a. Fit a simple linear regression model to predict the number of Facebook friends (Friends) as a function of the number of friend requests sent (Requests). Print a summary of the model.

b. What percent of the variation in number of friends can be explained by the model in the previous part?

c. Fit a multiple regression model to predict the number of friends (Friends) using the friend requests sent (Requests) and sex (Sex) as predictor variables. Print a summary of the model.

d. Create a scatterplot of Requests versus Friends. Overlay the scatterplot with the regression line for males as given in the previous model. Also, overlay the scatterplot with the regression line for females as given by the previous model.

e. How should the sex dummy variable be interpreted?

5. Facebook Analysis with Interactions

Using the Facebook.csv file, answer the following questions.

a. Fit a multiple regression model on the data set predicting Friends as a function of Sex and Requests. Include an interaction between the predictor variables. Print a summary of the model.

b. Interpret the interaction coefficient from the previous model.

c. Create a scatterplot of Requests versus Friends. Overlay the scatterplot with the regression line for males as given in the previous model. Also, overlay the scatterplot with the regression line for females as given by the previous model.

d. Using the multiple regression model with the interaction from part a, specify the simple linear regression equation which predicts the number of Friends using Requests for males. Also, specify the same simple linear regression equation for females.

e. Repeat the plot from the previous part and color the observations by Sex.

f. Fit a multiple regression model which uses the following predictors:

- A dummy variable for females
- The mean-centered friend requests
- An interaction

Chapter 8
Model Diagnostics

All generalizations are false, including this one.

—Mark Twain

8.1 Introduction

Knowing the correct mathematical model can be tremendously helpful when trying to make predictions. In many cases, however, a mathematical model that is relatively close to the true state may suffice. Yet this process brings its challenges. The quote from Mark Twain shown above alludes to the difficulty of attempting to find a mathematical model that is acceptable enough.

In previous chapters, we discussed how to find and evaluate models. In this chapter, we explore methods for evaluating regression models—one important tool is known as diagonostic plots—and determine if they are acceptable for a given situation. We will discuss diagnostic plots after a review of the mathematical model for multiple regression. While diagnostic plots greatly indicate when the regression model assumptions are violated, fixing these model violations requires additional analysis. We introduce methods for fixing some common model assumption violations. We also address unusual observations of the data that may occur in many cases. Lastly, a case study about an automotive sales conglomerate allows us to demonstrate how to run the analysis in detail with the relevant diagnostic plots and the R source code.

8.2 Multiple Regression Model Revisited

In Chap. 5, Eq. 5.1 describes how the response variable Y relates to the predictor variables $X_1, X_2, \ldots, X_p$ and an error term ε:

$$Y = \beta_0 + \beta_1 X_1 + \beta_2 X_2 + \cdots + \beta_p X_p + \varepsilon, \tag{8.1}$$

© The Author(s), under exclusive license to Springer Nature Switzerland AG 2023
D. P. McGibney, *Applied Linear Regression for Business Analytics with R*,
International Series in Operations Research & Management Science 337,
https://doi.org/10.1007/978-3-031-21480-6_8

where $\beta_0, \beta_1, \beta_2, \ldots, \beta_p$ are the model parameters. As alluded to in the chapter introduction, generalizations must be met with skepticism since the actual mathematical model can be elusive or even impossible to find. In Chap. 7, we explored the general linear model, which provides some techniques for modeling nonlinear relationships by manipulating the data to conform to the model from Eq. 5.1. After making such manipulations, it is important to determine if the model assumptions are met or if another model should be considered.

8.3   Model Assumptions

In Chap. 4, the assumptions for the simple linear regression model were stated. Here, we extend the assumptions to the multiple regression case. The four assumptions of the multiple regression model are listed below:

1. The relationship between the response and each predictor variable is linear (or $E[\varepsilon_i] = 0$ for each ε_i).
2. Each value of ε_i is independent.
3. The random variable, ε, is normally distributed.
4. The variance of ε, denoted as σ^2, is the same for all observations.

The assumptions are summarized in the statement below:

$$\varepsilon_i \overset{\text{i.i.d.}}{\sim} N(0, \sigma^2).$$

In this chapter, we discuss these assumptions in detail, as they are frequently violated in practice. The effect of the violation depends on which model assumption is not met and the degree of the divergence. In one such consequence, the coefficients might be biased, or some of the fundamental statistics from the regression may be invalid.

8.4   Violations of the Model Assumptions

If the relationship between a response and one or more predictor variables is nonlinear, then a linear model is not appropriate for the relationship. To resolve this issue, one can conduct transformations of the predictor variable as discussed in Chap. 7, which allows some nonlinear relationships to fit into the framework of the general linear model. In some cases, it may be necessary to transform the response variable.

If the errors (ε) are dependent, then the relationship between response and predictors is not correctly captured in the confidence and prediction intervals. For example, if the variance of the errors changes as the response increases, then the estimate of the standard deviation is invalid, which will make the confidence and prediction intervals invalid.

8.5 Residual Analysis

To check these linear regression assumptions, a number of diagnostic measures can be employed. One popular method is referred to as residual analysis. To analyze the residuals of a regression model, we can look at a plot of the residuals on the y-axis and the fitted values ($\hat{Y}$) on the x-axis. In the simple linear regression case, the predictor variable can be plotted on the x-axis in place of the fitted values.

Figure 8.1a shows a residual plot in which the residuals have roughly the same variation across all fitted values, from left to right. In addition, the slope of the residuals and the fitted values is zero. Figure 8.1b and d, on the other hand, demonstrates a quadratic or parabolic relationship. The residuals in Fig. 8.1c form a cubic relationship with the fitted values. Having a pattern such as a quadratic or cubic relationship as shown in the figures violates the linearity assumption.

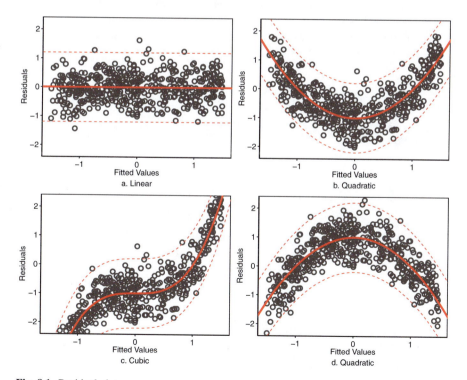

Fig. 8.1 Residual plots

8.6   Sales Performance Application: Residual Analysis

Recall from Chap. 7, we determined that a quadratic function best fit the sales data from Table 7.2. The linear model was fit using the following R commands.

```
df = read.csv("Employment.csv")
reg = lm(Sales ~ Employment, data = df)
```

From the simple linear regression model, do the following:

(a) Plot the residuals on the y-axis and the fitted values on the x-axis.
(b) Discuss the fit of the model based on the residual plot.
(c) What advantage does a residual plot have over a standard scatterplot between X and Y?

Solution

(a) The `plot` command can be used to plot the residuals as shown in Fig. 8.2. Specifically, from the plot on the left-side in the figure, the residuals are on the y-axis, and the fitted values on the x-axis. In base R, there are multiple plots generated by simply plotting a regression object. Here, we employ the `which` option to select the first of those plots. The result is shown in the right-side of Fig. 8.2. By plotting the regression object, a few additional details are shown, particularly the trend of the plot and the observation numbers of some values of interest.

```
par(mfrow = c(1, 2))
plot(reg$fitted.values, reg$residuals, xlab = "Fitted Values",
     ylab = "Residuals")
plot(reg, which = 1)
```

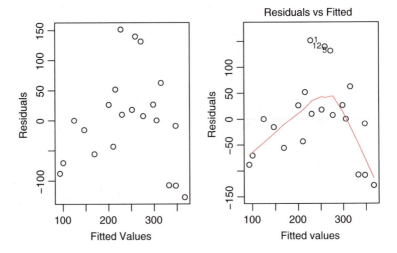

Fig. 8.2 Base R residual plot

(b) As indicated by the red curve, the residual plot suggests a nonlinear model and would benefit from quadratic modeling.
(c) When we analyzed this data set previously in Chap. 5, we found that the best fit for this data was a quadratic model. In this application, we arrive at the same conclusion using residual plots. In the residual plot from Fig. 8.2, the nonlinear relationship is exacerbated, which allows the analyst to more easily identify the best fit for the data.

8.7   Constant Variance

If a plot of the residuals against the predictor variable shows a pattern, then a strict requirement—called constant error variance—of the linear model is violated. The statistical term for constant error variance is homoscedasticity, whereas heteroscedasticity refers to nonconstant variance.

In Fig. 8.3a, no discernible pattern exists between the residuals and fitted values. The residuals have constant variance from left to right. The residuals from Fig. 8.3b increase in variation as the fitted values increase, resulting in a funnel-like shape as shown in Fig. 8.3b.

Identifying nonconstant variance from residual plots constitutes standard practice. For a more algorithmic procedure, a hypothesis test can be set up and concluded. This test, developed by Breush and Pagan in 1979, is referred to as the Breush–Pagan or the nonconstant variance test. The null and alternative hypotheses would then be

$$H_0 : \text{Constant variance}$$

$$H_1 : \text{Nonconstant variance.}$$

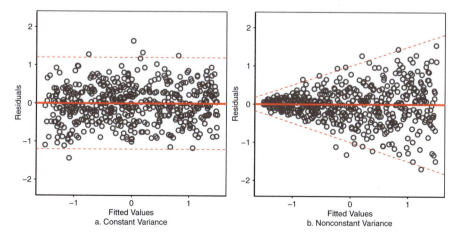

Fig. 8.3 Residual plots

In order to run a nonconstant variance test in R, the `ncvTest` function from the `car` package can be used. After loading in the `car` package and creating a model object `reg`, the `ncvTest` function can be used with the first argument being the model object. The syntax is given below.

```
ncvTest(reg)
```

This function calculates a chi-square test statistic and the corresponding p-value for the test. For the example above, the p-value generated for this test is 0.42557, which indicates that there is no evidence that the constant variance assumption has been violated.

8.8   Twitter Application: Residual Variance

Lina Alvarez aspires to be a social media influencer. To achieve her goal, she investigates the performance of social media influencers to understand the relationship between the number of tweets that an influencer posts per year and the followers added in the same year. Understanding this relationship will help her manage her own social media account with greater efficiency. She collects the number of followers (`Followers`) and the corresponding number of tweets (`Tweets`) from influencers in similar domains and saves the data in the `Twitter.csv` file.

Using the data in the `Twitter.csv` file located on the companion site, do the following:

(a) Load in the data and plot `Followers` on the y-axis and the `Tweets` on the x-axis. Comment on the relationship between the plotted variables.
(b) Fit a linear regression model that predicts `Followers` using `Tweets`. Use the model to fit a residual plot. Comment on the variance of the residuals.
(c) Test the residuals for nonconstant variance by performing a nonconstant variance test. Comment on the findings.

Solution

(a) A plot is easily generated by reading in the data and then using the `plot` command.

```
df = read.csv("Twitter.csv")
plot(Followers ~ Tweets, data = df)
```

While the `Followers` trends upward as `Tweets` increases, the variance of this trend increases as `Tweets` increases. This relationship can be seen by the upward trending funnel shape of the scatterplot in Fig. 8.4.

(b) After fitting the linear model using `lm`, we again use the `plot` command with the model as the first argument, followed by the `which` argument set to 1.

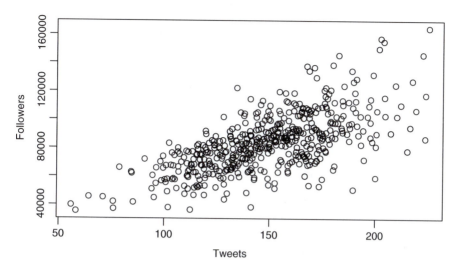

Fig. 8.4 Scatterplot of tweets vs. followers

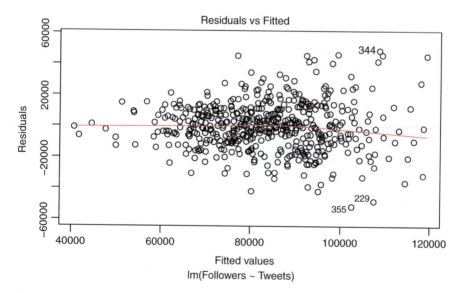

Fig. 8.5 Residual plot

```
reg = lm(Followers ~ Tweets, data = df)
plot(reg, which = 1)
```

From the output in Fig. 8.5, the increase of the residual variance becomes obvious from the funnel shape of the residual plot.

(c) Since the `ncvTest` is in the `car` package, use the `library` function to load in the `car` package. Then, use `ncvTest` with `reg` as the first argument.

```
library(car)
ncvTest(reg)
```

```
## Non-constant Variance Score Test
## Variance formula: ~ fitted.values
## Chisquare = 57.23239, Df = 1, p = 3.8724e-14
```

Here the p-value (3.8724×10^{-14}) is much less than $\alpha = 0.05$, which indicates that we reject the null hypothesis and conclude that the residuals have nonconstant variance.

8.9   Response Variable Transformations

In addition to predictor variable transformations, it may be helpful to transform the response variable. In fact, it is often the case that nonconstant variance can be corrected by transforming the response variable. The logarithmic transformation is a widely used transformation that we discuss here.

8.10   Logarithmic Transformations

In Fig. 8.6, the scatterplot on the left-side indicates exponential growth as X increases. We recognize the exponential growth by the changing increase in Y as X increases. Specifically, the values of Y increase at a slow rate when X is negative, but when X is positive, the points on the scatterplot increase significantly. Applying the logarithm of the response variable, Y, results in a linear relationship between $\log(Y)$ and X, as depicted on the right-side of Fig. 8.6.

The right-side of Fig. 8.7 exhibits a pattern of exponential decay. Noting the exponential decay, the logarithm of X is applied and plotted versus Y. The result is the linear relationship depicted on the left-side of Fig. 8.7.

In some data sets, it may be helpful to take the logarithm of both the response and predictor variables to model a linear relationship. On the left-side of Fig. 8.8, we note that many values appear bunched up between 0 and 2 for X and between 0 and 1 for Y. The logarithmic transform can "unbunch" these values. In the center of Fig. 8.8, we note that applying the logarithm of Y results in an exponential decay relationship. Therefore, the logarithm of X is applied, and the resulting relationship appears on the right-side of Fig. 8.8.

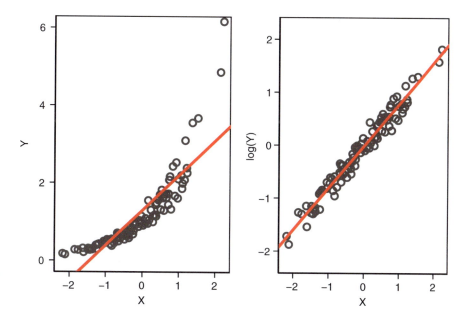

Fig. 8.6 Log(Y) transformation

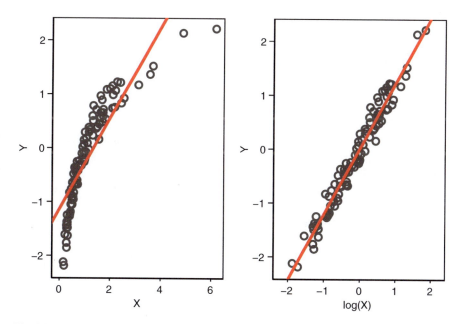

Fig. 8.7 Log(X) transformation

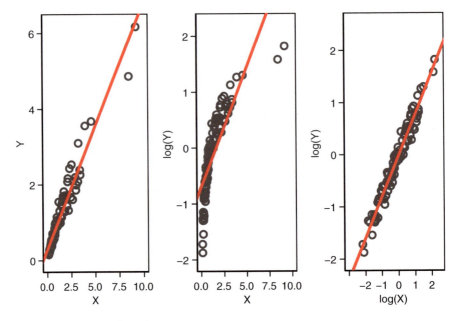

Fig. 8.8 Log–log transformation

8.11   Other Response Variable Transformations

Popular response variable transformations include the following:

- Logarithmic
- Reciprocal
- Square root

Of the three mentioned above, the logarithmic transformation is the most widely used. As an example of the reciprocal transformation, the left-side of Fig. 8.9 depicts a relationship between X and Y in which it is appropriate to transform Y with a reciprocal transformation. The action results in the model fit depicted in the right-side plot of Fig. 8.9.

In Fig. 8.10, the observed X and Y values are plotted along with the untransformed regression fit on the left-side figure. On the right-side of Fig. 8.10, the plot depicts the square root transformation of the response variable.

8.12   Box–Cox Transformation

Guidelines exist to help select the appropriate response variable transformation. One such set of guidelines is referred to as the Box–Cox method. Statisticians George E. P. Box and David R. Cox developed a procedure to identify the "optimal" power

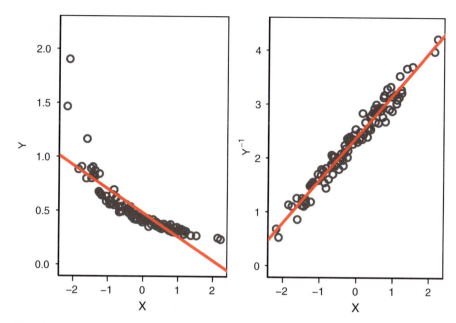

Fig. 8.9 Reciprocal transformation

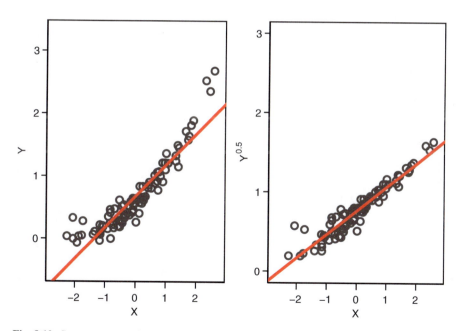

Fig. 8.10 Square root transformation

transformation of Y to stabilize the variance of the model residuals. In particular, the Box–Cox method finds the value of λ such that the value of Y^λ maximizes the log-likelihood function. Simply put, an analyst chooses the value of λ that is most likely to be the response variable exponent.

To use the Box–Cox method, we first find a model that predicts the untransformed Y. Then, the resulting model is passed along to the boxCox function. While other Box–Cox functions exist, we choose the implementation from the car package.

```
boxCox(reg)
```

After passing the regression model into the boxCox function, the log-likelihood values are on the y-axis, and values of λ are on the x-axis. Figure 8.11 shows an example of this plot.

While a more exact value of the estimate of λ can be found, an analyst commonly chooses the exponent of Y so that a common transformation is used. Therefore, we choose the exponent of Y to be one of the following:

- $\lambda = -1$ denotes the reciprocal transformation.
- $\lambda = 0$ denotes the logarithmic transformation.
- $\lambda = 0.5$ denotes the square root transformation.
- $\lambda = 1$ denotes that no transformation should be applied to Y.
- $\lambda = 2$ denotes a square transformation.

Figure 8.11 indicates that a logarithmic transformation would be most appropriate since $\lambda = 0$ is very close to the maximum log-likelihood.

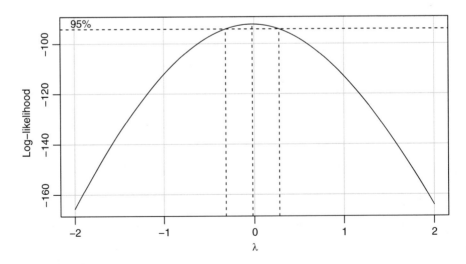

Fig. 8.11 Box–Cox results

8.13   Twitter Application: Box–Cox

Using the Twitter data from the previous application, use the techniques mentioned above to diagnose and rectify the nonconstant variance. Specifically, do the following:

(a) Apply the Box–Cox method to the linear model from the previous application and identify which transformation would be most appropriate.
(b) Fit a new model with the prescribed relationship and plot the residuals. Comment on the variance of the residuals.
(c) Perform a test for nonconstant variance on the revised model, and compare the resulting p-value with the previous application's p-value.

Solution

(a) The `reg` object from the previous application is used as the first argument of the `boxCox` function. If the `car` package was not previously loaded, it would be necessary to again use the `library` function in order to run `boxCox`.

```
library(car)
boxCox(reg)
```

From the Box–Cox results in Fig. 8.12, the value of λ that maximizes the log-likelihood function is close to 0.5. Therefore, we choose the response variable exponent of 0.5 to model the relationship between `Tweets` and `Followers`. Since taking a variable to the power of 0.5 is equivalent to taking the square root, this transformation is referred to as the square root transformation.

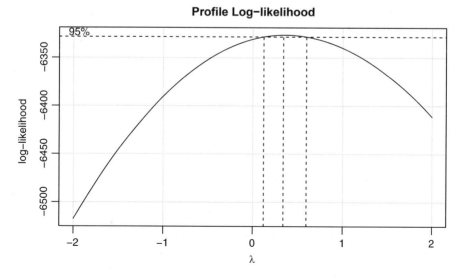

Fig. 8.12 Box–Cox results

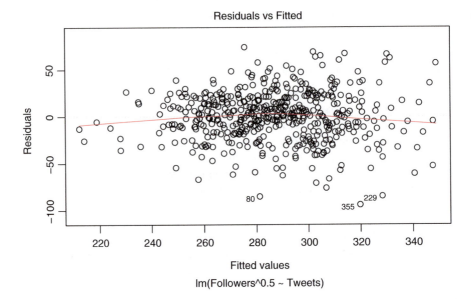

Fig. 8.13 Transformed response residual plot

(b) The new model is fit by changing the response in the formula to `Followers ^ .5`. After the transformed model is fit, we use the `plot` function to provide a residual plot.

```
Treg = lm(Followers ^ .5 ~ Tweets, data = df)
plot(Treg, which = 1)
```

In Fig. 8.13, there are only a few residuals with fitted values below 240. When the fitted values are above 240, the variance is relatively constant.

(c) Using the `ncvTest` function on the transformed regression model, we can conduct a test for nonconstant variance.

```
ncvTest(Treg)
```

```
## Non-constant Variance Score Test
## Variance formula: ~ fitted.values
## Chisquare = 22.84582, Df = 1, p = 1.7553e-06
```

While the p-value of the nonconstant variance test indicates that nonconstant variance is still evident, the p-value of 1.7553×10^{-06} indicates a significant improvement over the untransformed model with a p-value of 3.8724×10^{-14}.

8.14   Assessing Normality

Normality is often assessed by visually inspecting a histogram for a bell-shaped curve. Figure 8.14 shows a histogram with observed values of a continuous random variable X. The red curve indicates the pattern that a normal distribution would follow. Since the bars roughly align with the red curve, we would conclude that the distribution of X is approximately normal. This method of inspection is very intuitive since most people are familiar with the bell curve of a normal distribution.

Another popular method consists of plotting the ordered quantiles of the residuals on one axis versus the theoretical quantiles expected. This resulting plot is called a quantile–quantile plot (QQ plot). For normality to hold, the points should roughly fall on the line. The QQ plot in Fig. 8.15 is that of the random variable X from the histogram of Fig. 8.14.

While plots are generally sufficient for regression diagnostics, one might employ the Shapiro–Wilk hypothesis test as a test for normality. The null and alternative hypotheses would then be

$$H_0 : \text{Normal distribution}$$

$$H_1 : \text{Non-normal distribution.}$$

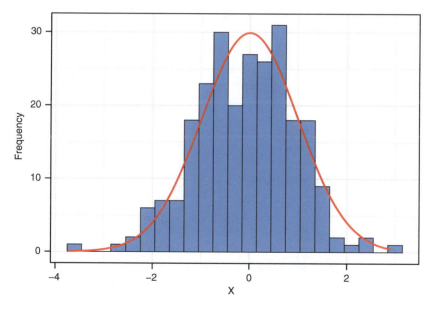

Fig. 8.14 Histogram

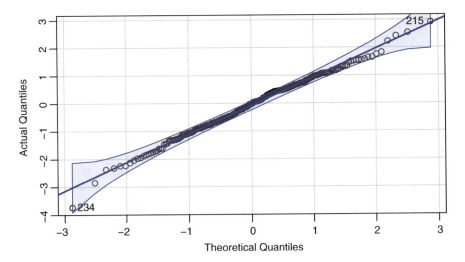

Fig. 8.15 Normal Q–Q plot

```
shapiro.test(X)
```

```
##
##   Shapiro-Wilk normality test
##
## data:   X
## W = 0.99469, p-value = 0.5344
```

The *p*-value of 0.5344 resulting from the Shapiro–Wilk test indicates that the
distribution of *X* is approximately normal, or at least that the null hypothesis should
not be rejected.

8.15   Assessing Independence

While tests for independence exist, the level of independence can be determined
by the study design or the collection of data. For instance, if observations are
ordered in a time sequence such as stock data, the current price of a stock is
dependent on the price yesterday. A common fix for this issue is to consider
a time series model. Another common problem occurs when observations are
clustered geographically, which would be better addressed using a spatial model.
When repeated measurements are drawn from the same subject, the resulting
errors are dependent. Repeated measurements can be regressed using differences
of measurements. Generally speaking, it is important to know the nature of the data
so that the proper model can be used to fit the data.

8.16   Outliers and Influential Observations

Some observations are more extreme than others. An extreme observation is referred to as an outlier, which can be revealed by plotting the residuals against a predictor variable. Outliers will be easy to identify because the residual will lie far away from the rest of the plot. If the outlier is influential, the results of the analysis may be affected. An influential observation refers to one that has a disproportionate effect on the value of the slope and intercept in the least squares regression equation.

Figure 8.16a shows a scatterplot with three extreme observations colored blue, red, and green. Figure 8.16b specifically isolates the blue observation from Fig. 8.16a with the black observations and two lines. The black line is fit using only the black observations, while the blue line is fit using the black observations plus the isolated blue observation. Note that the blue line does not differ significantly from the black line. Because the blue dot lies near the black dots, the black dots indicate that the Y value should be lower, showing only a slight difference between the blue and black lines.

Notice the red and green dots in Fig. 8.16a, c, and d; both red and green dots have an extreme X value compared to the other observations. However, the red dot lies near the black regression line, whereas the green dot lies far from it. Since the green dot is an outlier and no other observations occur close to it, the regression line changes significantly when the green observation is included, as indicated by the

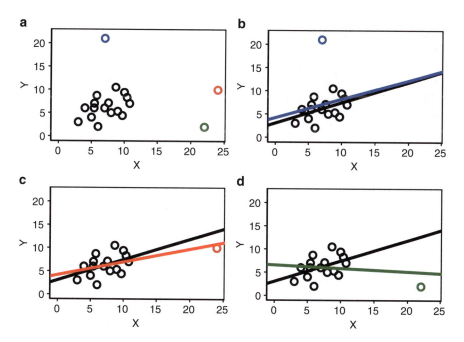

Fig. 8.16 Scatterplots illustrating the influence of outliers

green regression line. This drastic change to the regression line is in contrast to the red regression line, which remains somewhat similar to the black regression line. The green dot is considered an influential observation.

8.17   Residuals and Leverage

The leverage of an observation measures its ability to move the regression model all by itself by simply moving in the y direction. The leverage measures the amount by which the predicted value would change if the observation was shifted one unit in the y direction. The leverage always takes values between 0 and 1, inclusively. A point with zero leverage has no effect on the regression model. If a point has leverage equal to 1, the line must follow the point perfectly.

8.17.1   Leverage

Leverage can be attained for each predicted value in our data set. Taking the ith predicted value, we compose a linear combination of weights and the response value in the data set. For each predicted value, we have an equation of the form:

$$\hat{Y}_i = h_{i1}Y_1 + h_{i2}Y_2 + \ldots + h_{in}Y_n, \tag{8.2}$$

where the weights are denoted as h_{ij} and where i is the index of the predicted value and j corresponds to the weight from the corresponding observed value. From this relationship, we define the leverage of the ith value to be h_{ii}. If h_{ii} is large, then the y_i value has a large effect on $\hat{y}_i$, and we say that the ith value is highly leveraged.

Bringing together every value of h_{ij}, the hat matrix can be constructed, which is denoted as H.

```
hat(model.matrix(reg))
```

In the R code above, the model.matrix function takes the regression object and puts it into the correct variable type for the hat command.

8.17.2   Standardized Residuals

Prior to studying the residuals of the regression output, it is common practice to standardize them to compensate for differences in leverage. To standardize, one can simply divide by the standard deviation. The standardized residuals are given by

$$r_i = \frac{e_i}{s\sqrt{1 - h_{ii}}},$$

where e_i is the residual, s is the RMSE, and h_{ii} is the leverage for the ith observation.

8.17.3   Studentized Residuals

The standardized residuals are not valid t-values since e_i and s are dependent. Therefore, consider s_i, which is the RMSE with every residual except e_i. Similar to the standardized residuals, the studentized residuals are given as

$$r_i^* = \frac{e_i}{s_i \sqrt{1 - h_{ii}}}.$$

Studentized residuals can be computed in R using the command

```
r = rstudent(reg)
```

8.17.4   Cook's Distance

An influential point is one that if removed from the data, such point would significantly change the fit of the model. An influential point may either be an outlier or have large leverage, or both. Cook's distance refers to a commonly used influence measure that combines these two properties. It can be expressed as, "Typically, points with D_i greater than 1 are classified as being influential."

$$D_i = \left(\frac{r_i^2}{p} \right) \left(\frac{h_{ii}}{1 - h_{ii}} \right).$$

The construct may look a little messy, but the main aspect to recognize is that Cook's distance depends on both the residual, e_i, and the leverage, h_{ii}. That is, both the x value and the y value of the data point play a role in the calculation of Cook's distance.

In short:

- D_i directly summarizes how much all of the fitted values change when the ith observation is deleted.
- A data point having a large D_i indicates that the data point strongly influences the fitted values.

We can compute Cook's distance using the following command:

```
cook = cooks.distance(reg)
```

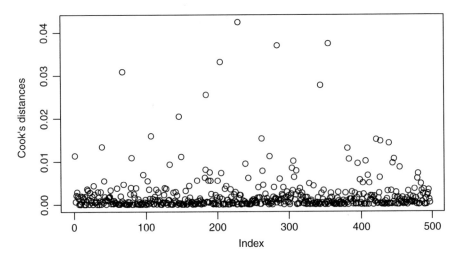

Fig. 8.17 Cook's distances for Twitter application

From the `cook` object created in the code above, a plot of the different Cook's distances can be generated using the plot command.

```
plot(cook, ylab="Cook's distances")
```

From Fig. 8.17, we see the largest Cook's distance for the Twitter Data analysis is about 0.0424 indicating that there are no significant influential points.

8.18 Case Study: Lead Generation

8.18.1 Problem Statement

Regression analysis has applications throughout many different business domains, including the evaluation of lead generation for sales revenue. Lead generation, or the methodology in which companies attract customers, results in sales and, therefore, generates revenue. For this reason, sales managers often work with analysts to find out which methods or tactics attract the most customers. Since data on customer acquisition are readily available, analysts serve a key role in understanding how to increase customer leads, thereby increasing sales revenue.

As a consulting analyst, you are tasked with helping Bergheger Automotive Group increase sales revenue. The group consists of 142 dealerships across the country. Management is interested in understanding their lead generation results from 2021. You request data on the various lead generation methods so that you can analyze and make recommendations. The sales team currently has several methods of customer acquisition available to them: radio ads, robocalls, email, and cold-calls.

Each method carries its own strengths and weaknesses. In this analysis, you need to understand the relationship between sales revenue and the corresponding customer acquisition method. Specifically, your tasks consist of the following:

1. Identify revenue by lead generation method.
2. Find the dealerships with the highest total revenue.
3. Describe the relationship between revenue and each lead generation method by fitting a simple linear regression line and interpreting the fitted model.
4. Make recommendations on how to increase revenue.

8.18.2   Data Description

You are given the Sales.csv file that contains data from 142 dealerships with the following variables:

- X1 is the number of radio ads.
- Y1 is the sales revenue generated from the radio ads in dollars.
- X2 is the number of robocalls in hundreds.
- Y2 is the sales revenue generated from robocalls in dollars.
- X3 is the number of emails in thousands.
- Y3 is the sales revenue generated from emails in dollars.
- X4 is the number of cold-calls in hundreds.
- Y4 is the sales revenue generated from cold-calls in dollars.

As a first task, we will do some simple descriptive statistics to understand the data. After loading in the data, a summary of the variables is calculated using the summary function.

```
df = read.csv("Sales.csv")
summary(df)
```

```
##        X1               Y1                X2
##   Min.   : 16.00   Min.   : 15457   Min.   : 14.11
##   1st Qu.: 83.25   1st Qu.: 47530   1st Qu.: 51.12
##   Median :118.00   Median : 84490   Median : 88.77
##   Mean   :119.53   Mean   :121145   Mean   :124.32
##   3rd Qu.:158.00   3rd Qu.:165425   3rd Qu.:170.16
##   Max.   :236.00   Max.   :740712   Max.   :880.47
##        Y2               X3                Y3
##   Min.   : 6294    Min.   :  4.498   Min.   : 15674
##   1st Qu.:33402    1st Qu.: 17.443   1st Qu.: 47222
##   Median :47140    Median : 34.700   Median : 84123
##   Mean   :47823    Mean   : 54.224   Mean   :120833
##   3rd Qu.:63214    3rd Qu.: 77.444   3rd Qu.:166332
##   Max.   :94410    Max.   :368.485   Max.   :721275
```

```
##         X4                 Y4
## Min.    :35.67   Min.    : 57519
## 1st Qu.:50.11    1st Qu.: 69128
## Median :65.32    Median : 74500
## Mean    :66.33   Mean    : 75959
## 3rd Qu.:80.75    3rd Qu.: 81189
## Max.    :99.27   Max.    :111973
```

The summary indicates that the largest value among response variables is $740,712, which can be observed by looking at the maximum value of Y1. The Y1 response variable, which corresponds to radio ad revenue, also shows the largest mean value. The mean and maximum values of Y1 are followed closely by the Y3 response variable, corresponding to email revenue. Revenue from cold-calls, represented by Y4, has the third lowest mean followed by robocalls, represented by Y2.

In addition to viewing the summary function, an analyst may benefit from looking at some of the observations. As discussed previously, the head function will display the first six observations.

```
head(df)
```

```
##   X1     Y1      X2     Y2     X3      Y3    X4     Y4
## 1 126  72705  56.33 50375 40.757  75617 89.67  80251
## 2 116 110808 134.15 46315 33.269 107596 37.21  71193
## 3  56  35567  37.33 22573 10.151  35303 57.29  62825
## 4  95  74220  85.25 38017 21.971  72655 90.82  88127
## 5 140 138520 148.41 55891 53.701 137058 89.84  86198
## 6 144 138982 140.84 57796 59.067 138699 94.47 111973
```

Observing the head of the data provides some intuition about the data. For example, a reasonable value of X1 is 100 radio ads that would most likely generate revenue in the low one hundred thousand. Therefore, the revenue generated by an ad is roughly just above $1 thousand. While this is a rough approximation, it can help understand the meaning of the X1 variable and its relationship to Y1. Rough estimates of the other variable relationships can similarly aid in understanding the data.

8.18.3 Revenue by Lead Generation Method

Notice that the amount of revenue from the various sources is shown in the barplot in Fig. 8.18. The revenue variables are summed individually using the sum function and combined into a vector using the c function. By using the names function, a name is attached to each sum. The barplot function displays the specified name on the x-axis of the barplot. In addition to specifying the ylab and main arguments,

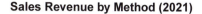

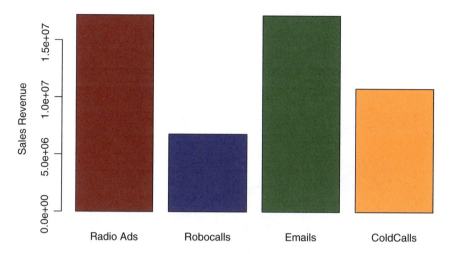

Fig. 8.18 Barplot of sales revenue by method

the `col` argument is used to specify different colors for each revenue generation method. The order of the colors specified corresponds to the order of the sums. In particular, radio ads will be dark red, robocalls will be dark blue, emails will be dark green, and cold-calls will be dark orange. These colors were chosen since they are aesthetically pleasing and for easy viewing. We will adhere to this color coding in later plots throughout this case.

```
sums = c(sum(df$Y1), sum(df$Y2), sum(df$Y3), sum(df$Y4))
names(sums) = c("Radio Ads", "Robocalls", "Emails", "ColdCalls")
barplot(sums, ylab = "Sales Revenue",
        main = "Sales Revenue by Method (2021)",
        col = c("darkred", "darkblue", "darkgreen",
                "darkorange"))
```

Figure 8.18 supports our assessment of the response variables from observing the summary output. Specifically, the revenue from radio ads and emails is the highest, followed by the revenue from cold-calls and robocalls. Note that the largest total revenue for each response variable will have the same proportional value as the mean of said response variable. In fact, after dividing the sales revenue by the number of dealerships, each revenue variable will retain the same ratio as depicted in Fig. 8.18.

While the `barplot` function provides some basic functionality, the `ggplot2` package provides more flexibility and less limitations. For a more thorough discussion on `ggplot`, see Chap. 9.

Boxplots of the different revenue amounts can be calculated using the `boxplot` function. As a first attempt, we select the revenue variables by using the `subset` function. Rather than using the variable names Y1, Y2, Y3, and Y4, we use the `names`

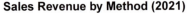

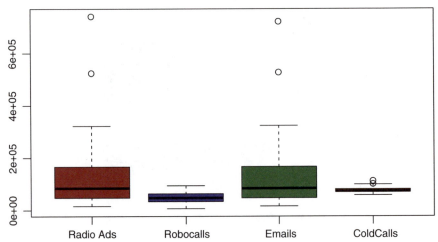

Fig. 8.19 Boxplot of sales revenue by method

function to modify the names of the data subset. We then pass the data subset to the `boxplot` function and specify the `col` argument to change the color of each respective box in the plot.

```
sub = subset(df, select = c(Y1, Y2, Y3, Y4))
names(sub) = c("Radio Ads", "Robocalls", "Emails", "ColdCalls")
boxplot(sub, main = "Sales Revenue by Method (2021)",
        col = c("darkred", "darkblue", "darkgreen",
                "darkorange"))
```

In Fig. 8.19, the variation of the radio ad and email variables can be observed. However, the variation of the robocalls and cold-calls is difficult to decipher since the y-axis is dictated by the revenue range from the radio ads and emails. As a simple remedy, we can plot the boxplot of radio ads and emails together using the same y-axis scale and then plot the boxplot of robocalls and cold-calls on a separate y-axis scale. The result is shown in Fig. 8.20.

We use the `subset` function to select the responses, creating the data sets: `sub1` and `sub2`. The `sub1` dataframe contains radio ad and email revenue variables, while the `sub2` dataframe contains the robocall and cold-call revenue variables. After generating the data subsets, the use of the `names` and `boxplot` functions is similar to the previous boxplot. Notice the different y-axis for the left-side and right-side boxplots.

Revenue by Radio Ads, Emails   **Revenue by Robocalls, Cold−calls**

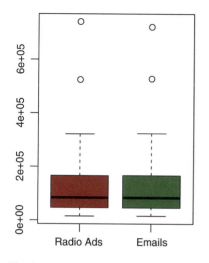 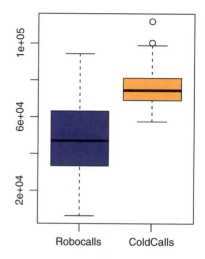

Fig. 8.20 Boxplots of sales revenue by method

```
par(mfrow=c(1,2))

# Ads and Emails
sub1 = subset(df, select = c(Y1, Y3))
names(sub1) = c("Radio Ads", "Emails")
boxplot(sub1, col = c("darkred", "darkgreen"),
        main = "Revenue by Radio Ads, Emails")

# Robocalls and cold-calls
sub2 = subset(df, select = c(Y2, Y4))
names(sub2) = c("Robocalls", "ColdCalls")
boxplot(sub2, col = c("darkblue","darkorange"),
        main = "Revenue by Robocalls, Cold-calls")
```

On the left-side of Fig. 8.20, the revenue by radio ads and emails shows two similar boxplots, whereas the right-side shows that revenue by robocalls is not very similar to that of cold-calls. For most dealerships, the revenue from cold-calls will be larger than the revenue for robocalls since the bottom whisker of cold-calls lands above the median line for robocalls.

8.18.4   *Revenue by Dealership*

To calculate the total revenue for each dealership, all lead sources for each individual dealership are summed. We will name this column `Total` and sort it from highest to lowest. Sorting a column from highest to lowest is referred to as sorting in descending order. To sort the data in descending order, the sort function can be used by specifying the column to be sorted, followed by the decreasing argument set to TRUE. Omitting this decreasing argument would default to sorting the column in increasing order or ascending order. Similar to the sort function, the order function returns the row number corresponding to the sorted totals.

```
Total = df$Y1 + df$Y2 + df$Y3 + df$Y4
sorted_rev = sort(Total, decreasing = TRUE)
sorted_idx = order(Total, decreasing = TRUE)
```

The names function is used to specify the dealership identification number and is referred to as sorted_idx. A barplot of the top five revenues by dealership number can be found by specifying the first five values of the sorted_rev vector as the first argument in the barplot function.

```
names(sorted_rev) = sorted_idx
barplot(sorted_rev[1:5], xlab = "Dealership Number",
        ylab = "Revenue", main = "Revenue by Dealership")
```

From the barplot in Fig. 8.21, we see that dealership 88 has the highest total revenue in 2021, followed by dealerships 54, 80, 131, and 100. While stacked barplots can be used to show the breakdown of revenue by ad source for each of the top five dealerships, this task is performed better using the ggplot2 package. For more details, the dealerships with the top radio ad, robocall, email, and cold-call

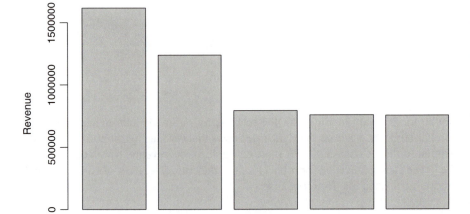

Fig. 8.21 Barplot of sales revenue by dealership

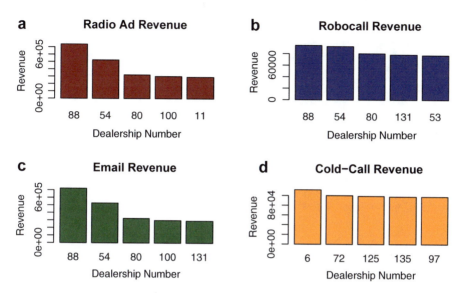

Fig. 8.22 Revenue type by dealership

revenue are shown in Fig. 8.22. Each subfigure in Fig. 8.22 is generated by sorting the respective revenue variables in ascending order and returning the corresponding dealership number as done in the previous figure (Fig. 8.21). Note that dealerships 88, 54, and 80 are consistently the highest in every revenue category with the exception of the cold-calls. The dealerships with the top cold-calls revenue are 6, 72, 125, 135, and 97.

```
par(mfrow=c(2,2))

# Ad Revenue
sorted_ad_rev = sort(df$Y1, decreasing = TRUE)
names(sorted_ad_rev) = order(df$Y1, decreasing = TRUE)
barplot(sorted_ad_rev[1:5], xlab = "Dealership Number",
        ylab = "Revenue", main = "Radio Ad Revenue",
        col = "darkred")

# Robocall Revenue
sorted_rc_rev = sort(df$Y2, decreasing = TRUE)
names(sorted_rc_rev) = order(df$Y2, decreasing = TRUE)
barplot(sorted_rc_rev[1:5], xlab = "Dealership Number",
        ylab = "Revenue", main = "Robocall Revenue",
        col = "darkblue")

# Email Revenue
```

```
sorted_email_rev = sort(df$Y3, decreasing = TRUE)
names(sorted_email_rev) = order(df$Y3, decreasing = TRUE)
barplot(sorted_email_rev[1:5], xlab = "Dealership Number",
        ylab = "Revenue", main = "Email Revenue",
        col = "darkgreen")

# Cold-Call Revenue
sorted_cc_rev = sort(df$Y4, decreasing = TRUE)
names(sorted_cc_rev) = order(df$Y4, decreasing = TRUE)
barplot(sorted_cc_rev[1:5], xlab = "Dealership Number",
        ylab = "Revenue", main = "Cold-Call Revenue",
        col = "darkorange")
```

8.18.5 Sales Versus Radio Ads

The relationship between sales and radio ads is clearly nonlinear as indicated in
Fig. 8.23. It can be argued that the trend is parabolic; however, a much better fit
would be that of an exponential function. To account for this exponential, we can
apply a logarithmic transform onto the Y1 variable, since applying a logarithmic
transformation to Y1 with X1 as a predictor is equivalent to predicting Y1 with $e =
2.718\ldots$ raised to the power of X1 for positive values of Y1.

```
par(mfrow = c(1,2))
plot(Y1 ~ X1, data = df, xlab = "Ads", ylab = "Sales",
```

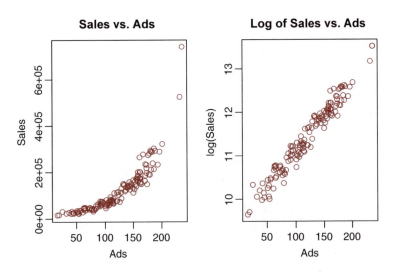

Fig. 8.23 Scatterplots of sales and ads (left) and transformed sales and ads (right)

```
          main = "Sales vs. Ads", col = "darkred")
plot(log(Y1) ~ X1, data = df, xlab = "Ads", ylab = "log(Sales)",
          main = "Log of Sales vs. Ads", col = "darkred")
```

In Fig. 8.23, the fit of the second scatterplot follows a linear pattern, which indicates a fit of the logarithmic transformation is appropriate using the `lm` function. The summary of the transformed model is obtained by passing the `reg` object to the `summary` function.

```
reg = lm(log(Y1) ~ X1, data = df)
summary(reg)
```

```
##
## Call:
## lm(formula = log(Y1) ~ X1, data = df)
##
## Residuals:
##       Min       1Q    Median       3Q      Max
## -0.39550 -0.11163  0.01475  0.11820  0.43745
##
## Coefficients:
##                  Estimate Std. Error t value Pr(>|t|)
## (Intercept) 9.4877598  0.0389673  243.48   <2e-16 ***
## X1          0.0161339  0.0003035   53.16   <2e-16 ***
## ---
## Signif. codes:
## 0 '***' 0.001 '**' 0.01 '*' 0.05 '.' 0.1 ' ' 1
##
## Residual standard error: 0.1695 on 140 degrees of freedom
## Multiple R-squared:  0.9528, Adjusted R-squared:  0.9525
## F-statistic:  2826 on 1 and 140 DF,  p-value: < 2.2e-16
```

After fitting the transformed model, we inspect the residuals using a histogram and a residual plot. The histogram is generated using the `hist` function. Note that the residuals are within the regression object (`reg`) that allows us to specify `reg$residuals` as the first argument of the `hist` function.

```
par(mfrow = c(1, 2))
hist(reg$residuals, xlab = "Residuals", main = "Histogram",
          col = "darkred")
plot(reg$fitted.values, reg$residuals, xlab = "Fitted values",
          ylab = "Residuals", main = "Residual Plot",
          col = "darkred")
```

The diagnostic plots indicate the normality condition is met within reason, and the variance of the residuals does not violate any assumptions. More specifically,

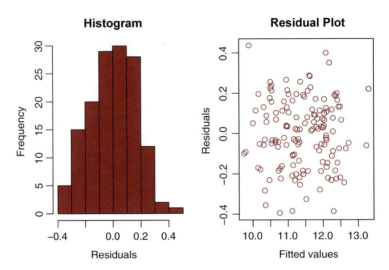

Fig. 8.24 Diagnostic plots of transformed ads model

the histogram from the left-side of Fig. 8.24 roughly exhibits a bell-shaped normal distribution, and the right-side residual plot does not appear to show a pattern, hence meeting the linearity condition. Since the conditions are met, the coefficients can be interpreted. Here, the coefficient is 0.0161339, which indicates that for every additional radio ad, the logarithm of revenue increases by 0.0161339. Alternatively, the revenue gets multiplied by a product of $e^{0.0161339} = 1.016265$ for every additional radio ad. This amounts to a 1.6265% increase in the revenue per radio ad. Therefore, if the revenue is high, the increase will be more significant. This multiplicative increase may be due to the compounding effect of having multiple ads.

8.18.6 Sales Versus Robocalls

From Fig. 8.25, we see that the relationship between robocalls and sales is also nonlinear. In this case, the relationship can be modified by transforming the X2 variable with a logarithmic transform.

```
par(mfrow = c(1,2))
plot(Y2 ~ X2, data = df, xlab = "Robocalls (hundreds)",
     ylab = "Sales", main = "Sales vs. Robocalls",
     col = "darkblue")
plot(Y2 ~ log(X2), data = df, xlab = "log(Robocalls (hundreds))",
     ylab = "Sales", main = "Sales vs. Log of Robocalls",
     col = "darkblue")
```

```
reg = lm(Y2 ~ log(X2), data = df)
summary(reg)
```

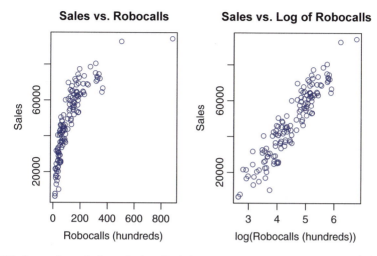

Fig. 8.25 Scatterplots of sales and robocalls (left) and sales and transformed robocalls (right)

```
##
## Call:
## lm(formula = Y2 ~ log(X2), data = df)
##
## Residuals:
##     Min      1Q Median      3Q     Max
## -21417   -4194    -933    4880   17698
##
## Coefficients:
##                Estimate Std. Error t value Pr(>|t|)
## (Intercept)  -48407.2     3238.4  -14.95   <2e-16 ***
## log(X2)       21379.3      707.9   30.20   <2e-16 ***
## ---
## Signif. codes:
## 0 '***' 0.001 '**' 0.01 '*' 0.05 '.' 0.1 ' ' 1
##
## Residual standard error: 6889 on 140 degrees of freedom
## Multiple R-squared:  0.8669, Adjusted R-squared:  0.866
## F-statistic: 912.1 on 1 and 140 DF,  p-value: < 2.2e-16
```

After fitting the revised model, we inspect the residuals using a histogram and a residual plot. The generated plots are shown in Fig. 8.26.

```
par(mfrow = c(1,2))
hist(reg$residuals, xlab = "Residuals", main = "Histogram",
     col = "darkblue")
plot(reg$fitted.values, reg$residuals, xlab = "Fitted values",
```

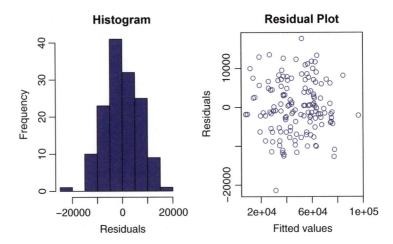

Fig. 8.26 Diagnostic plots for transformed robocall model

```
ylab = "Residuals", main = "Residual Plot",
col = "darkblue")
```

The diagnostic plots indicate that the residuals are normally distributed, and the model is linear with constant error variance. Given that the diagnostic plots indicate the linear regression model assumptions are met and the p-value of the coefficient is low, the coefficients can be interpreted. The coefficient of the log of robocalls is 21,379.3, indicating for every unit increase in the logarithm of robocalls, there will be an increase in revenue of $21,379.30. The alternative interpretation holds that for every unit increase in the exponent of robocalls, the sales revenue increases by $21,379.30. As demonstrated in the untransformed plot, the initial increase may be steep, but the returns diminish quickly. With robocalls, repeated calls are often ignored, so it is logical to use the robocalls sparingly, unless there is no penalty for making frequent calls.

8.18.7 Sales Versus Emails

The left-side of Fig. 8.27 shows the untransformed scatterplot of sales and emails. Note the observations are densely distributed around the lower values of the x-axis and become more dispersed as x increases. Similarly, the observations are densely distributed around the lower values of the y-axis and become more dispersed as y increases. This pattern indicates each axis can benefit from a logarithmic transformation. Applying logarithms to both x and y axes changes the pattern in the left-side scatterplot to a linear relationship, which can be observed on the right-side of Fig. 8.27.

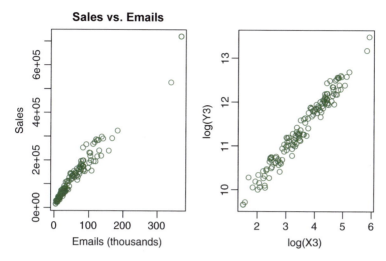

Fig. 8.27 Scatterplots of sales and emails (left) and transformed sales and transformed emails (right)

```
par(mfrow = c(1, 2))
plot(Y3 ~ X3, data = df, xlab = "Emails (thousands)",
     ylab = "Sales", main = "Sales vs. Emails",
     col = "darkgreen")
plot(log(Y3) ~ log(X3), data = df, col = "darkgreen")
```

Using the lm function, the transformed model is fit. Since we named the regression object as reg, the previous regression object is replaced with this new fit. The summarized model is shown by using the summary function.

```
reg = lm(log(Y3) ~ log(X3), data = df)
summary(reg)
```

```
##
## Call:
## lm(formula = log(Y3) ~ log(X3), data = df)
##
## Residuals:
##      Min       1Q   Median       3Q      Max
## -0.35257 -0.10007  0.01142  0.10735  0.38332
##
## Coefficients:
##               Estimate Std. Error t value Pr(>|t|)
## (Intercept)    8.53364    0.04969  171.74   <2e-16 ***
## log(X3)        0.80550    0.01343   59.99   <2e-16 ***
```

```
## ---
## Signif. codes:
## 0 '***' 0.001 '**' 0.01 '*' 0.05 '.' 0.1 ' ' 1
##
## Residual standard error: 0.15 on 140 degrees of freedom
## Multiple R-squared:  0.9626, Adjusted R-squared:   0.9623
## F-statistic:   3599 on 1 and 140 DF,  p-value: < 2.2e-16
```

After fitting the revised model, we inspect the residuals using a histogram and a residual plot, as done previously.

```
par(mfrow = c(1, 2))
hist(reg$residuals, xlab = "Residuals", main = "Histogram",
     col = "darkgreen")
plot(reg$fitted.values, reg$residuals, xlab = "Fitted values",
     ylab = "Residuals", main = "Residual Plot",
     col = "darkgreen")
```

The diagnostic plots in Fig. 8.28 indicate normally distributed residuals with constant variance and a linear pattern; therefore, the linear regression assumptions are met. Furthermore, the coefficient for log of emails is significant, indicating the coefficient can be interpreted. For a log–log model, the slope can be interpreted as a 1% increase in emails is associated with a 0.8055% increase in revenue.

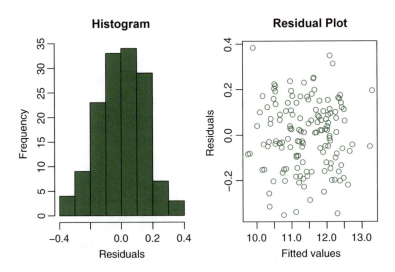

Fig. 8.28 Diagnostic plots of transformed email model

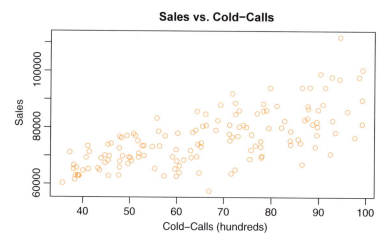

Fig. 8.29 Scatterplot of sales and cold-calls

8.18.8 Sales Versus Cold-Calls

Observing the scatterplot of sales versus cold-calls in Fig. 8.29, the plot appears to have nonconstant variance since the variation increases from left to right. This increase in variation takes on the shape of a funnel.

```
plot(Y4 ~ X4, data = df, xlab = "Cold-Calls (hundreds)",
     ylab = "Sales", main = "Sales vs. Cold-Calls",
     col = "darkorange")
```

To verify that the plot has nonconstant variance, we fit the simple linear regression model and generate the diagnostic plots for the model.

```
reg = lm(Y4 ~ X4, data = df)
#summary(reg)
```

Inspecting the residuals using a histogram and a residual plot is done easily using the code here.

```
par(mfrow = c(1, 2))
hist(reg$residuals, xlab = "Residuals", main = "Histogram",
     col = "darkorange")
plot(reg$fitted.values, reg$residuals, xlab = "Fitted values",
     ylab = "Residuals", main = "Residual Plot",
     col = "darkorange")
```

Note that the residual plot in Fig. 8.30 clearly shows nonconstant variance since the spread in the residuals increases from left to right. The increasing variance

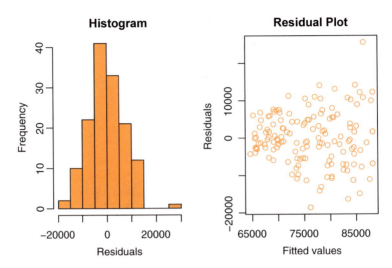

Fig. 8.30 Diagnostic plots of cold-calls

pattern is that of a funnel. While it is clear that nonconstant variance is present, a simple hypothesis test can be performed in order to verify the results. Within the car package, the ncvTest function tests for nonconstant variance using a chi-square test. If the car package is not installed, the install.packages function should be run in order to install it.

```
library(car)
ncvTest(reg)
```

```
## Non-constant Variance Score Test
## Variance formula: ~ fitted.values
## Chisquare = 21.63455, Df = 1, p = 3.2986e-06
```

Running the nonconstant variance test indicates indeed there is nonconstant variance since the p-value from the test is extremely low. To remedy this condition, we will apply the Box–Cox transform using the boxCox function, which is also from the car package. Note that the library function only needs to run once to access both the ncvTest and boxCox functions.

```
boxCox(reg)
```

From the Box–Cox output in Fig. 8.31, we observe that the exponent of the response variable (λ) maximizes the log-likelihood function when $\lambda = 1$. Therefore, we take revenue to the power of -1 as the response variable here. The transformed response becomes 1/Y4.

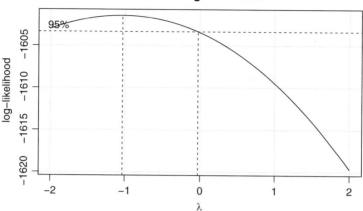

Fig. 8.31 Box–Cox output

```
reg = lm(1/Y4 ~ X4, data = df)
summary(reg)
```

```
##
## Call:
## lm(formula = 1/Y4 ~ X4, data = df)
##
## Residuals:
##         Min        1Q     Median         3Q        Max
## -2.692e-06 -9.549e-07 -2.000e-10  8.439e-07  4.050e-06
##
## Coefficients:
##               Estimate Std. Error t value Pr(>|t|)
## (Intercept)  1.748e-05  3.775e-07   46.31   <2e-16 ***
## X4          -6.202e-08  5.493e-09  -11.29   <2e-16 ***
## ---
## Signif. codes:
## 0 '***' 0.001 '**' 0.01 '*' 0.05 '.' 0.1 ' ' 1
##
## Residual standard error: 1.176e-06 on 140 degrees of
## freedom
## Multiple R-squared:  0.4766, Adjusted R-squared:  0.4729
## F-statistic: 127.5 on 1 and 140 DF,  p-value: < 2.2e-16
```

After fitting the revised model, we inspect the residuals using a histogram and a residual plot to check the model assumptions.

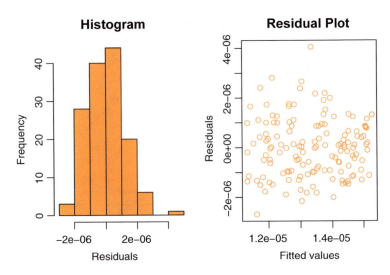

Fig. 8.32 Diagnostic plots for transformed cold-calls model

```
par(mfrow = c(1, 2))
hist(reg$residuals, xlab = "Residuals", main = "Histogram",
    col = "darkorange")
plot(reg$fitted.values, reg$residuals, xlab = "Fitted values",
    ylab = "Residuals", main = "Residual Plot",
    col = "darkorange")
```

The diagnostic plots in Fig. 8.32 indicate the linear regression assumptions are somewhat violated. Specifically, the histogram indicates non-normality given its distribution, and the residual plot indicates there is still evidence of nonconstant variance. While the residuals appear to be slightly normally distributed, the shapiro.test function can be used to assess normality with a hypothesis test.

```
shapiro.test(reg$residuals)
```

```
##
##  Shapiro-Wilk normality test
##
## data:  reg$residuals
## W = 0.98461, p-value = 0.1129
```

Since the p-value of the Shapiro–Wilk test is not more extreme than $\alpha = 0.05$, the test indicates the residual distribution is normal. While the nonconstant variance appears to have improved significantly from the revised residual plot, there appears to be traces of nonconstant variance. The ncvTest is used here on the revised model.

```
ncvTest(reg)
```

```
## Non-constant Variance Score Test
## Variance formula: ~ fitted.values
## Chisquare = 4.093229, Df = 1, p = 0.043055
```

Note the p-value of the revised model drastically increased from the untransformed model. While the revised p-value is still below $\alpha = 0.05$, the nonconstant variance test may actually be better suited for lower values of α.

Here, we interpret the coefficient for the transformed model since the conditions of linear regression are nearly met, and the p-value of the coefficient indicates significance. The coefficient indicates that for every 100 calls, there is a $6.202e - 08$ decrease in the inverse of revenue.

8.18.9 Recommendations and Findings

From this analysis come multiple insights. In particular, regarding the tasks that were given, we find the following:

1. Identify revenue by lead generation method.

   Task 1 was answered through the use of careful observation and plots, which made clear that ads and emails were the most effective in generating revenue, while cold-calls also provided some benefit, followed by robocalls.

2. Find the dealerships with the highest total revenue.

   Plots of revenue by dealership in Figs. 8.21 and 8.22 were used to answer the second task. Dealerships 88, 54, and 80 are the highest in every revenue category except the cold-call revenue. The dealerships with the top cold-call revenue are 6, 72, 125, 135, and 97.

3. Describe the relationship between revenue and each lead generation method by fitting a simple linear regression line and interpreting the fitted model.

   Each regression model was fit and interpreted. Some interpretations, however, were more practical than others. The interpretations are:

   - The ad revenue increases by a multiple of 1.6265% for every additional ad.
   - For every unit increase in the exponent of robocalls, the sales revenue increases by $21,379.30.
   - For a 1% increase in emails, there is a 0.8055% increase in revenue.
   - For every 100 call increase in the number of cold-calls, there is a 6.202e-08 decrease in the inverse of revenue from cold-calls.

4. Make recommendations on how to increase revenue.

- From observing the dealerships by revenue, dealership 88 can provide insights to other dealerships to increase revenue from ads, email, and robocalls.
- Also, dealership 6 has the most effective means by which to generate revenue from cold-calls, which could be helpful to share with other dealerships.
- Since ads increase revenue exponentially, dealerships should focus on increasing the number of ads, assuming that ads are not cost-prohibitive.
- While robocalls may be effective in the short run, if there is a cost associated with having extra calls, then the number of robocalls should be limited. Also, the negative publicity of robocalls should be considered when determining the number of calls to make.
- Both emails and cold-calls are beneficial methods of lead generation and should be continued if they are not cost-prohibitive.

8.18.10 Case Conclusion

This analysis demonstrates how studying marketing data can lead to increasing revenue streams for a car dealership conglomerate. Analyzing marketing methods can help optimize profits and other performance metrics. Analytics is becoming more relevant in the field of marketing as data become more abundant.

Real-world data, such as marketing data, can be messy and require variable transformations and diagnostic measures to better understand the relationship between responses and predictors. In this case, the responses required transformations in order to adhere to the assumptions of linear regression. We used many base R commands to model and analyze the data, such as: c, plot, boxplot, barplot, lm, data.frame, hist, subset, summary, order, sort, and shapiro.test. In addition to the base R commands, we used the ncvTest and boxCox functions that are within the car package.

Problems

1. **Sales from Web Ads Diagnostics**
   Using the SalesAds.csv file, answer the following questions:

   a. Fit a simple linear regression model to predict the sales in thousands of USD (Sales) as a function of the number of website ads (Ads) without using transformations. Print a summary of the model.
   b. Generate a scatterplot of Sales and Ads. Include the linear regression line in the plot.
   c. Construct a scatterplot and histogram of the residuals for the linear model from part a.
   d. From the scatterplot and histogram in part c, which model assumptions appear to be violated?

e. Determine an appropriate transformation for correcting the problems found in part d and fit the corresponding regression model.

f. Plot the transformed data and the regression line from part e.

g. Construct a scatterplot and histogram of the residuals for the linear model from part a.

2. Health Care Customers Diagnostics

Using the CustomersInsurance.csv file, answer the following questions:

a. Fit a simple linear regression model to predict the number of customers per day (Customers) as a function of the number of people with Aetna insurance plans (Plans) within a 5 mile radius of each clinic without using transformations. Print a summary of the model.

b. Generate a scatterplot of Customers and Plans. Include the linear regression line in the plot.

c. Construct a scatterplot and histogram of the residuals for the linear model from part a.

d. From the scatterplot and histogram in part c, which model assumptions appear to be violated?

e. Since the observations appear to be bunched up around the x-axis, apply a logarithmic transformation to the x-axis and generate a model summary.

f. Plot the transformed data and the regression line from part e.

g. Construct a scatterplot and histogram of the residuals for the linear model from part e.

3. House Price Prediction Diagnostics

Using the PriceSize.csv file, answer the following questions:

a. Fit a simple linear regression model to predict the price in thousands of USD (Price) as a function of the size of the house (Size) in square feet without using transformations. Print a summary of the model.

b. Generate a scatterplot of Price and Size. Include the linear regression line in the plot.

c. Construct a scatterplot and histogram of the residuals for the linear model from part a.

d. From the scatterplot and histogram in part c, which model assumptions appear to be violated?

e. Since the observations appear to be bunched up around the x and y axes, apply logarithmic transformations to both X and Y. Generate a model summary.

f. Plot the transformed data and the regression line from part e.

g. Construct a scatterplot and histogram of the residuals for the linear model from part e.

Chapter 9
Variable Selection

All models are wrong, but some are useful.

—George E. P. Box

9.1 Introduction

While it may be a bit strict to say that *all* models are wrong, it is often the case that a model is imperfect. However, an imperfect model may still provide a great amount of value. When attempting to find the best model from the data given, being able to select the predictor variables is of utmost importance in the model building process. In fact, one of the most important aspects of model creation is knowing which predictor variables to use, a process sometimes called feature selection or variable selection. Variable selection can be tremendously helpful when an analyst is attempting to find a mathematical model that is relatively close to the true state.

In this chapter, a methodology for finding a subset of predictor variables to create an acceptable model is explored. Particularly, iterative approaches for finding a good model are explored along with the best subset selection method. We also discuss model evaluation criteria to assist the analyst in selecting a model. To further motivate the concepts discussed in this chapter, an Airbnb data set is used. Finally, the chapter case study consists of finding an appropriate model to predict health care costs using an insurance data set.

9.2 Parsimonious Models

The major goals of regression analysis are to have a model that makes accurate predictions and also can be easily interpreted. These attributes describe a parsimonious model. In order to have a parsimonious model, we should find the model with the fewest number of predictors that still gives a relatively good model fit.

© The Author(s), under exclusive license to Springer Nature Switzerland AG 2023
D. P. McGibney, *Applied Linear Regression for Business Analytics with R*,
International Series in Operations Research & Management Science 337,
https://doi.org/10.1007/978-3-031-21480-6_9

9.3   Airbnb Pricing Application

Victoria would like to list her property on Airbnb in Edinburgh, the capitol of
Scotland. In order to price her property competitively, she collects data on listings
to analyze the contributing factors of price. The AirBnb.csv data set consists of
10,370 rental listings from Airbnb in Edinburgh for a period from June 25, 2019 to
June 24, 2020.

The variables for each listing are:

- Bathrooms—Number of bathrooms
- Bedrooms—Number of bedrooms
- Beds—Number of beds
- Accommodates—Number of guests the listing can accommodate
- Guests—Number of guests included without an additional fee
- MinNights—Minimum number of nights required for booking
- MaxNights—Maximum number of nights the listing can be rented
- ExtraPeople—Average fee for each additional person in British pounds
- HostListings—Number of listings the host manages
- ResponseRate—Average host response rate
- Deposit—Average security deposit required for booking in British pounds
- CleaningFee—Average cleaning fee charged in British pounds
- FeeMissing—A dummy variable that is 1 if the cleaning fee is missing, 0
  otherwise
- Price—Average price of the listing in British pounds.

Using the Edinburgh Airbnb data file, do the following:

(a) Load in the data and print the first six observation to the console.
(b) Fit a regression model predicting Price using all other variables and print the
    model summary.
(c) Repeat the previous part, but exclude the predictor variable with the lowest
    individual significance.
(d) Comment on the model fit for the models in parts (b) and (c). Which model is
    more parsimonious?

Solution

(a) Using the read.csv function, the data frame df is created. The first six
    observations are printed to the console by passing df to the head function.

```
df = read.csv("AirBnb.csv")
head(df)
```

```
##    Bathrooms Bedrooms Beds Accommodates Guests MinNights
## 1       1.0        1    1            2      1       2.9
## 2       1.5        2    2            4      4       2.0
## 3       1.0        0    2            2      2       4.2
```

```
## 4          1.0        1   1            2     2      2.0
## 5          1.0        1   1            2     2      1.0
## 6          1.0        1   2            3     2      1.0
##    MaxNights ExtraPeople HostListings ResponseRate Deposit
## 1        30           0            1               1     200
## 2       365          25            1               1     250
## 3        60          10            2               1     100
## 4        21           0            1               1       0
## 5        31           0            1               1      75
## 6        31          10            2               1       0
##    CleaningFee FeeMissing      Price
## 1         40.0          0  101.05479
## 2         30.0          0  111.81644
## 3         15.4          1   49.98356
## 4         10.0          0   33.82740
## 5         10.0          0   80.24110
## 6          8.0          0   77.65753
```

(b) The reg1 model is fit using the lm function. We specify that Price is predicted
     using all other variables in the data frame using the formula as the first
     argument. The model summary is printed using the summary function.

```
reg1 = lm(Price ~ ., data = df)
summary(reg1)
```

```
##
## Call:
## lm(formula = Price ~ ., data = df)
##
## Residuals:
##      Min       1Q    Median       3Q      Max
## -131.302  -24.807   -8.201   15.966  248.684
##
## Coefficients:
##                 Estimate Std. Error t value Pr(>|t|)
## (Intercept)    1.576e+01  3.805e+00    4.140 3.49e-05 ***
## Bathrooms      1.266e+01  1.032e+00   12.258  < 2e-16 ***
## Bedrooms       8.551e+00  9.413e-01    9.084  < 2e-16 ***
## Beds           3.300e+00  6.704e-01    4.922 8.70e-07 ***
## Accommodates   1.288e+01  5.216e-01   24.698  < 2e-16 ***
## Guests        -3.618e-01  3.742e-01   -0.967 0.333641
## MinNights      2.015e+00  3.517e-01    5.730 1.03e-08 ***
## MaxNights     -1.495e-03  7.393e-04   -2.023 0.043124 *
## ExtraPeople   -1.667e-01  3.590e-02   -4.643 3.47e-06 ***
```

```
## HostListings   3.977e-01   2.477e-02   16.054   < 2e-16 ***
## ResponseRate  -1.295e+01   3.579e+00   -3.619 0.000298 ***
## Deposit        6.053e-02   5.223e-03   11.589   < 2e-16 ***
## CleaningFee     8.095e-02   2.354e-02    3.440 0.000585 ***
## FeeMissing      2.164e+00   9.393e-01    2.304 0.021259 *
## ---
## Signif. codes:
## 0 '***' 0.001 '**' 0.01 '*' 0.05 '.' 0.1 ' ' 1
##
## Residual standard error: 40.32 on 10356 degrees of freedom
## Multiple R-squared:  0.4764, Adjusted R-squared:  0.4758
## F-statistic: 724.9 on 13 and 10356 DF,  p-value: < 2.2e-16
```

(c) From the model summary in the previous part, we note that all predictor variables are significant with the exception of Guests which has an individual significant *p*-value of 0.333641. Here we fit reg2 by specifying a formula that includes all of the predictor variables except the Guests variable.

```
reg2 = lm(Price ~ . - Guests, data = df)
summary(reg2)
```

```
##
## Call:
## lm(formula = Price ~ . - Guests, data = df)
##
## Residuals:
##      Min       1Q   Median       3Q      Max
## -130.615  -24.697   -8.165   15.920  248.739
##
## Coefficients:
##                 Estimate Std. Error t value Pr(>|t|)
## (Intercept)    1.557e+01  3.800e+00    4.097 4.23e-05 ***
## Bathrooms      1.266e+01  1.032e+00   12.259   < 2e-16 ***
## Bedrooms       8.548e+00  9.413e-01    9.081   < 2e-16 ***
## Beds           3.267e+00  6.696e-01    4.879 1.08e-06 ***
## Accommodates   1.280e+01  5.144e-01   24.880   < 2e-16 ***
## MinNights      2.038e+00  3.509e-01    5.806 6.58e-09 ***
## MaxNights     -1.491e-03  7.393e-04   -2.016 0.043784 *
## ExtraPeople   -1.779e-01  3.399e-02   -5.233 1.70e-07 ***
## HostListings   3.939e-01  2.447e-02   16.101   < 2e-16 ***
## ResponseRate  -1.296e+01  3.579e+00   -3.621 0.000294 ***
## Deposit        6.070e-02  5.220e-03   11.629   < 2e-16 ***
## CleaningFee     7.809e-02  2.335e-02    3.345 0.000827 ***
## FeeMissing      2.249e+00  9.352e-01    2.404 0.016213 *
## ---
```

```
## Signif. codes:
## 0 '***' 0.001 '**' 0.01 '*' 0.05 '.' 0.1 ' ' 1
##
## Residual standard error: 40.32 on 10357 degrees of freedom
## Multiple R-squared:  0.4764, Adjusted R-squared:  0.4758
## F-statistic: 785.3 on 12 and 10357 DF,  p-value: < 2.2e-16
```

(d) While the multiple R^2 and adjusted R^2 values are relatively unchanged, the reg2 model is more parsimonious since it has fewer predictors.

9.4 Assessing Model Performance

We take into consideration several metrics to assess our models. Each metric assesses different aspects about model performance and so we discuss the details of some of the most important metrics here.

9.4.1 Multiple R-Squared and Adjusted R-Squared

In Chap. 3, the R^2 value was introduced for simple linear regression, and later, in Chap. 5, the multiple R^2 and adjusted R^2 were introduced and built off of the knowledge of R^2. Naturally, the multiple R^2 represents the amount of explained variation over the total variation,

$$R^2 = SSR/SST. \tag{9.1}$$

However, it was noted that the explained variation is incorrectly increased by irrelevant variables. To counteract this problem, we take into consideration the number of predictor variables (p) and the number of observations (n):

$$R_a^2 = 1 - (1 - R^2)\frac{n-1}{n-p-1}, \tag{9.2}$$

which gives the adjusted R^2 value. Both the multiple R^2 and the adjusted R^2 are given from the model summary within R.

9.4.2 Akaike Information Criterion

One commonly used criterion for model selection is the Akaike Information Criterion (AIC). This criterion takes into account the SSE, the number of observations (n), and the number of predictors (p),

$$AIC = n \log \frac{SSE}{n} + 2p. \tag{9.3}$$

To use the AIC criterion, we find AIC values for each model we would like to compare and then choose the model with the lower AIC value. The AIC value can be calculated in R using the `AIC` function.

9.4.3   Bayesian Information Criterion

Another commonly used criterion for model selection is the Bayesian Information Criterion (BIC). The equation for BIC is similar to Eq. 9.3,

$$BIC = n \log \frac{SSE}{n} + p \log n. \tag{9.4}$$

Note that while AIC uses $2p$ as the last component of the equation, BIC uses $p \log n$. This subtle difference will increase BIC more rapidly for an increase in the number of predictors (p). Thus, BIC is penalized more heavily for higher values of p resulting in fewer predictor variables in the chosen model. The BIC value can be calculated in R using the `BIC` function.

9.4.4   Mallows's C_p

One additional measure is Mallows's C_p. This measure is also popular for model selection. This measure is calculated by

$$C_p = \frac{SSE}{MSE_{full}} - n + 2p, \tag{9.5}$$

where SSE is calculated from the model in question and MSE_{full} is the mean squared error calculated from the model with all considered predictors. Like the previous criteria, Mallows's C_p can be used to find a good model with few variables. While we seek to find the smallest value of C_p, we also deem the model to be good if $C_p \leq p$. One should note that Mallows's C_p should only be used if you are not considering using the full model as your final model. For the full model, it can easily be proven that $C_p = p$, so Mallows's statistic will always deem the full model good.

9.5   Airbnb Pricing Application: Model Comparison

The models from the previous application were fit using the following code:

```
reg1 = lm(Price ~ ., data = df)
reg2 = lm(Price ~ . - Guests, data = df)
```

Using the models from the previous application, answer the following:

(a) How do we account for the difference between the multiple R^2 and the adjusted R^2 in the first model? Is this difference relatively large or small?
(b) Calculate the AIC and BIC for the first model.
(c) Calculate the AIC and BIC for the second model.
(d) Which model has the lower AIC? Which model has the lower BIC?

Solution

(a) The adjusted R^2 takes into account the number of predictors and the number of observations. There is a very small difference between the multiple R^2 and the adjusted R^2 in the first model.
(b) The AIC and BIC values are calculated by passing the reg1 model to the AIC and BIC functions.

```
AIC(reg1)
BIC(reg1)
```

```
## [1] 106118.4
```

```
## [1] 106227.1
```

(c) Here we repeat the calculations from the previous part using reg2.

```
AIC(reg2)
BIC(reg2)
```

```
## [1] 106117.4
```

```
## [1] 106218.8
```

(d) The reg2 model has a lower AIC and a lower BIC indicating that reg2 is a better model than reg1.

9.6   Backward Elimination

In the first application of this chapter, we fit a model using all of the predictor variables in the data set, and then we used a model without the least significant predictor variable. In essence we were doing backward elimination with the individual significance as the selection criteria. However, most backward elimination algorithms use AIC, BIC, or adjusted R^2 as the selection criteria.

Backward elimination begins with a model that includes all predictor variables the analyst would like to consider. After fitting this model with all of the considered

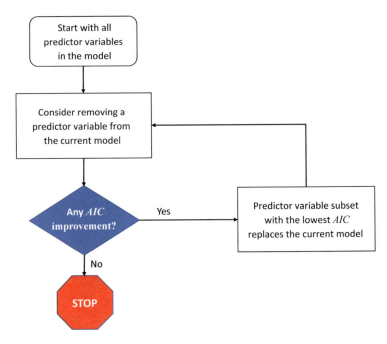

Fig. 9.1 Backward elimination

variables, the analyst excludes one variable, which results in the best model selection metric. The process of excluding variables one at a time is repeated until no improvement can be made in the model selection metric. From this process, it should be noted that a predictor variable cannot reenter at a subsequent step or iteration. A structural outline of this algorithm is depicted in Fig. 9.1.

The `step` function is an implementation of backward elimination in R. The first argument of `step` should be the model with all predictor variables under consideration. If the model with all predictors is defined as `full`, backward elimination can be performed using the R code:

```
step(full)
```

The `step` function begins with a full model and considers removing one predictor variable at a time. The AIC is calculated for each model with one variable removed, and the model chosen for the next iteration is the one with the lowest AIC. The algorithm is repeated on this reduced model. The algorithm continues in this fashion and stops when removing a variable does not reduce the AIC. While not specified, the `direction` argument is set to the default `"backward"` for backward elimination. One thing to note is that the AIC calculated in the `step` function is a simplification of the true AIC which is fine for the comparison of models, but it differs mathematically from Eq. 9.3 and the R function `AIC`.

9.7 Airbnb Pricing Application: Backward Elimination

Using the Edinburgh Airbnb data file, do the following:

(a) Fit a linear regression model using all possible predictor variables.
(b) Run backward elimination beginning from the full model in the previous part.
(c) Specify which variable is eliminated in the first iteration of backward elimination. How many variables are eliminated in total using this process?
(d) Print a model summary of the backward elimination model.

Solution

(a) The model with all the predictor variables was found previously as `reg1`. Here
we fit the model again naming the resulting model object as `full`.

```
full = lm(Price ~ ., data = df)
```

(b) Backward elimination is run using the `step` function. The `full` model is the
only necessary argument to run backward elimination using this function.

```
BE = step(full)
```

```
## Start:  AIC=76687.66
## Price ~ Bathrooms + Bedrooms + Beds + Accommodates +
##      Guests + MinNights + MaxNights + ExtraPeople +
##      HostListings + ResponseRate + Deposit + CleaningFee +
##      FeeMissing
##
##                 Df Sum of Sq      RSS    AIC
## - Guests         1      1520 16838997 76687
## <none>                       16837477 76688
## - MaxNights      1      6652 16844129 76690
## - FeeMissing     1      8629 16846105 76691
## - CleaningFee    1     19235 16856711 76697
## - ResponseRate   1     21290 16858766 76699
## - ExtraPeople    1     35052 16872529 76707
## - Beds           1     39388 16876864 76710
## - MinNights      1     53374 16890851 76718
## - Bedrooms       1    134171 16971648 76768
## - Deposit        1    218375 17055852 76819
## - Bathrooms      1    244304 17081781 76835
## - HostListings   1    419010 17256487 76941
## - Accommodates   1    991776 17829253 77279
##
## Step:  AIC=76686.59
## Price ~ Bathrooms + Bedrooms + Beds + Accommodates +
```

```
##      MinNights + MaxNights + ExtraPeople + HostListings +
##      ResponseRate + Deposit + CleaningFee + FeeMissing
##
##                  Df Sum of Sq       RSS   AIC
## <none>                        16838997 76687
## - MaxNights      1       6610 16845607 76689
## - FeeMissing     1       9400 16848396 76690
## - CleaningFee    1      18188 16857184 76696
## - ResponseRate   1      21323 16860319 76698
## - Beds           1      38710 16877707 76708
## - ExtraPeople    1      44522 16883519 76712
## - MinNights      1      54812 16893809 76718
## - Bedrooms       1     134084 16973080 76767
## - Deposit        1     219853 17058850 76819
## - Bathrooms      1     244326 17083322 76834
## - HostListings   1     421491 17260487 76941
## - Accommodates   1    1006421 17845418 77287
```

(c) The Guests variable is eliminated in the first iteration. The Guests variable is the only variable eliminated.

(d) Here we pass the BE model to the summary function.

```
summary(BE)
```

```
##
## Call:
## lm(formula = Price ~ Bathrooms + Bedrooms + Beds +
##      Accommodates + MinNights + MaxNights + ExtraPeople +
##      HostListings + ResponseRate + Deposit + CleaningFee +
##      FeeMissing, data = df)
##
## Residuals:
##      Min       1Q   Median       3Q      Max
## -130.615  -24.697   -8.165   15.920  248.739
##
## Coefficients:
##               Estimate Std. Error t value Pr(>|t|)
## (Intercept)   1.557e+01  3.800e+00    4.097 4.23e-05 ***
## Bathrooms     1.266e+01  1.032e+00   12.259  < 2e-16 ***
## Bedrooms      8.548e+00  9.413e-01    9.081  < 2e-16 ***
## Beds          3.267e+00  6.696e-01    4.879 1.08e-06 ***
## Accommodates  1.280e+01  5.144e-01   24.880  < 2e-16 ***
## MinNights     2.038e+00  3.509e-01    5.806 6.58e-09 ***
## MaxNights    -1.491e-03  7.393e-04   -2.016 0.043784 *
```

```
## ExtraPeople  -1.779e-01  3.399e-02  -5.233 1.70e-07 ***
## HostListings  3.939e-01  2.447e-02  16.101  < 2e-16 ***
## ResponseRate -1.296e+01  3.579e+00  -3.621 0.000294 ***
## Deposit       6.070e-02  5.220e-03  11.629  < 2e-16 ***
## CleaningFee   7.809e-02  2.335e-02   3.345 0.000827 ***
## FeeMissing    2.249e+00  9.352e-01   2.404 0.016213 *
## ---
## Signif. codes:
## 0 '***' 0.001 '**' 0.01 '*' 0.05 '.' 0.1 ' ' 1
##
## Residual standard error: 40.32 on 10357 degrees of freedom
## Multiple R-squared:  0.4764, Adjusted R-squared:  0.4758
## F-statistic: 785.3 on 12 and 10357 DF,  p-value: < 2.2e-16
```

9.8 Forward Selection

Forward selection begins without any predictor variables in the model. This model without predictor variables is referred to as the null model and only contains the intercept. Beginning with this null model, we consider all possible models with one predictor variable. Once we choose the best model, we continue adding in predictor variables, one at a time, until there is no longer an improvement in the selection criterion. In this case, we are bringing variables into the model, and we do not consider removing them in the forward selection algorithm. A structural outline of this algorithm is depicted in Fig. 9.2.

To implement forward selection in R, we use the step function and specify the direction and scope arguments. Particularly, if we have a null model, null, and a full model, full, then we code the following:

```
step(null, scope = list(upper = full), direction = "forward")
```

To specify the model consisting of all variables under consideration as the upper model, the syntax in the given code necessitates the use of the list function.

9.9 Airbnb Pricing Application: Forward Selection

Using the Edinburgh Airbnb data file, do the following:

(a) Fit a linear regression model using only the intercept.
(b) Run forward selection beginning from the model in the previous part.
(c) Specify which variable is incorporated in the first iteration of forward selection.
(d) Print a model summary of the forward selection model.

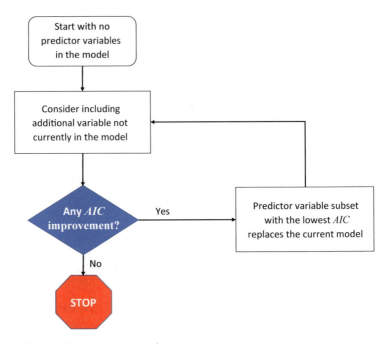

Fig. 9.2 Forward selection

Solution

(a) In an R formula, 1 represents the intercept. Thus, the null model is fit using the formula: Price ~ 1.

```
null = lm(Price ~ 1, data = df)
```

(b) The step function is used to perform forward selection by beginning with the null model and considering all the predictor variables from the full model. The scope and direction arguments are also necessary inputs to perform forward selection. The complete syntax is specified here to create the FS object. Here we print only the first 2 iterations for brevity.

```
FS = step(null, scope = list(upper = full), direction = "forward")
```

```
## Start:  AIC=83372.16
## Price ~ 1
##
##                   Df Sum of Sq       RSS    AIC
## + Accommodates     1  13670568  18489144  77634
## + Bedrooms         1  11685570  20474142  78692
## + Beds             1  11109178  21050534  78979
```

```
## + CleaningFee     1    4836817 27322895 81684
## + Bathrooms       1    4619579 27540133 81766
## + Guests          1    3196247 28963465 82289
## + Deposit         1    1772373 30387339 82786
## + HostListings    1    1190895 30968817 82983
## + MinNights       1     970397 31189315 83056
## + FeeMissing      1     403422 31756290 83243
## + ExtraPeople     1      21360 32138352 83367
## + MaxNights       1      20803 32138909 83367
## <none>                        32159712 83372
## + ResponseRate    1        155 32159557 83374
##
## Step:  AIC=77634.05
## Price ~ Accommodates
##
##                   Df Sum of Sq      RSS   AIC
## + Bathrooms        1    465453 18023690 77372
## + Bedrooms         1    439845 18049299 77386
## + HostListings     1    415354 18073790 77400
## + Deposit          1    318547 18170597 77456
## + MinNights        1    222259 18266885 77511
## + Beds             1    118146 18370998 77570
## + ExtraPeople      1     94287 18394857 77583
## + CleaningFee      1     79041 18410103 77592
## + FeeMissing       1     45143 18444001 77611
## + ResponseRate     1     36707 18452437 77615
## <none>                        18489144 77634
## + Guests           1      2967 18486176 77634
## + MaxNights        1      1844 18487300 77635
##
## Step:  AIC=77371.65
## Price ~ Accommodates + Bathrooms
```

(c) The first variable incorporated into the model is Accommodates. From the beginning of the forward selection output, we note that the starting AIC takes on a value of 83372.16. Note that +Accommodates denotes the inclusion of this variable and the corresponding AIC of 77634 is the lowest AIC on the first iteration. Therefore, including Accommodates is the best improvement on the model at the first iteration.

(d) The FS model is easily printed using the summary function.

```
summary(FS)
```

```
##
## Call:
## lm(formula = Price ~ Accommodates + Bathrooms +
##     HostListings + Deposit + Bedrooms + MinNights +
##     ExtraPeople + Beds + ResponseRate + CleaningFee +
##     FeeMissing + MaxNights, data = df)
##
## Residuals:
##      Min      1Q   Median      3Q      Max
## -130.615  -24.697   -8.165   15.920  248.739
##
## Coefficients:
##                  Estimate Std. Error t value Pr(>|t|)
## (Intercept)    1.557e+01  3.800e+00    4.097 4.23e-05 ***
## Accommodates   1.280e+01  5.144e-01   24.880  < 2e-16 ***
## Bathrooms      1.266e+01  1.032e+00   12.259  < 2e-16 ***
## HostListings   3.939e-01  2.447e-02   16.101  < 2e-16 ***
## Deposit        6.070e-02  5.220e-03   11.629  < 2e-16 ***
## Bedrooms       8.548e+00  9.413e-01    9.081  < 2e-16 ***
## MinNights      2.038e+00  3.509e-01    5.806 6.58e-09 ***
## ExtraPeople   -1.779e-01  3.399e-02   -5.233 1.70e-07 ***
## Beds           3.267e+00  6.696e-01    4.879 1.08e-06 ***
## ResponseRate  -1.296e+01  3.579e+00   -3.621 0.000294 ***
## CleaningFee    7.809e-02  2.335e-02    3.345 0.000827 ***
## FeeMissing     2.249e+00  9.352e-01    2.404 0.016213 *
## MaxNights     -1.491e-03  7.393e-04   -2.016 0.043784 *
## ---
## Signif. codes:
## 0 '***' 0.001 '**' 0.01 '*' 0.05 '.' 0.1 ' ' 1
##
## Residual standard error: 40.32 on 10357 degrees of freedom
## Multiple R-squared:  0.4764, Adjusted R-squared:  0.4758
## F-statistic: 785.3 on 12 and 10357 DF,  p-value: < 2.2e-16
```

9.10 Stepwise Regression

While stepwise regression could possibly refer to backward elimination or forward selection, we will use it here to refer to the more general case of stepwise regression where predictor variables could be added or removed from the model. The stepwise algorithm starts with a null model and then not only adds in variables, one at a time,

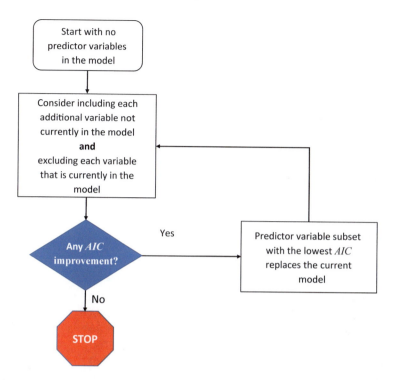

Fig. 9.3 Stepwise regression

but also considers removing variables from the model. The stepwise method differs from the forward selection method because we consider removing variables from the model as well. A structural outline of this algorithm is depicted in Fig. 9.3.

Since the stepwise algorithm is similar to forward selection, the code to implement stepwise is similar to that of forward selection. Particularly, if we have null and full models defined as `null` and `full`, respectively, the `step` function can be used to invoke the stepwise algorithm:

```
SW = step(null, scope = list(upper = full), direction = "both")
```

Notice the `direction` argument is specified as `"both"` to allow variables to be added or removed in each step.

9.11 Airbnb Pricing Application: Stepwise Regression

Using the Edinburgh Airbnb data file, do the following:

(a) Run stepwise regression beginning from a model using only the intercept.
(b) Specify which variable is incorporated in the first iteration of the stepwise regression.

(c) Print a model summary of the stepwise regression model.

(d) Clearly state the difference (if any) between the backwards elimination, forward selection, and the stepwise regression models.

Solution

(a) Using the `null` and `full` models as inputs, stepwise regression is performed using the `step` function. Note that the syntax is identical to that of the forward selection method with the exception of the `direction` argument which is specified as `"both"`. Here we print only the first 2 iterations for brevity.

```
SW = step(null, scope = list(upper = full), direction = "both")
```

```
## Start:  AIC=83372.16
## Price ~ 1
##
##                    Df Sum of Sq      RSS   AIC
## + Accommodates      1  13670568 18489144 77634
## + Bedrooms          1  11685570 20474142 78692
## + Beds              1  11109178 21050534 78979
## + CleaningFee       1   4836817 27322895 81684
## + Bathrooms         1   4619579 27540133 81766
## + Guests            1   3196247 28963465 82289
## + Deposit           1   1772373 30387339 82786
## + HostListings      1   1190895 30968817 82983
## + MinNights         1    970397 31189315 83056
## + FeeMissing        1    403422 31756290 83243
## + ExtraPeople       1     21360 32138352 83367
## + MaxNights         1     20803 32138909 83367
## <none>                          32159712 83372
## + ResponseRate      1       155 32159557 83374
##
## Step:  AIC=77634.05
## Price ~ Accommodates
##
##                    Df Sum of Sq      RSS   AIC
## + Bathrooms         1    465453 18023690 77372
## + Bedrooms          1    439845 18049299 77386
## + HostListings      1    415354 18073790 77400
## + Deposit           1    318547 18170597 77456
## + MinNights         1    222259 18266885 77511
## + Beds              1    118146 18370998 77570
## + ExtraPeople       1     94287 18394857 77583
## + CleaningFee       1     79041 18410103 77592
## + FeeMissing        1     45143 18444001 77611
## + ResponseRate      1     36707 18452437 77615
```

```
## <none>                              18489144 77634
## + Guests         1          2967 18486176 77634
## + MaxNights      1          1844 18487300 77635
## - Accommodates   1  13670568 32159712 83372
##
## Step:  AIC=77371.65
## Price ~ Accommodates + Bathrooms
```

(b) The `Accommodates` variable is incorporated in the first iteration. Note that on the second iteration, the stepwise method considers eliminating the `Accommodates` variable from the model even though that would result in a large increase in the AIC.

(c) A summary of the `SW` object is done here.

```
summary(SW)
```

```
##
## Call:
## lm(formula = Price ~ Accommodates + Bathrooms +
##      HostListings + Deposit + Bedrooms + MinNights +
##      ExtraPeople + Beds + ResponseRate + CleaningFee +
##      FeeMissing + MaxNights, data = df)
##
## Residuals:
##       Min        1Q    Median        3Q       Max
## -130.615   -24.697    -8.165    15.920   248.739
##
## Coefficients:
##                  Estimate Std. Error t value Pr(>|t|)
## (Intercept)     1.557e+01  3.800e+00   4.097 4.23e-05 ***
## Accommodates    1.280e+01  5.144e-01  24.880  < 2e-16 ***
## Bathrooms       1.266e+01  1.032e+00  12.259  < 2e-16 ***
## HostListings    3.939e-01  2.447e-02  16.101  < 2e-16 ***
## Deposit         6.070e-02  5.220e-03  11.629  < 2e-16 ***
## Bedrooms        8.548e+00  9.413e-01   9.081  < 2e-16 ***
## MinNights       2.038e+00  3.509e-01   5.806 6.58e-09 ***
## ExtraPeople    -1.779e-01  3.399e-02  -5.233 1.70e-07 ***
## Beds            3.267e+00  6.696e-01   4.879 1.08e-06 ***
## ResponseRate   -1.296e+01  3.579e+00  -3.621 0.000294 ***
## CleaningFee     7.809e-02  2.335e-02   3.345 0.000827 ***
## FeeMissing      2.249e+00  9.352e-01   2.404 0.016213 *
## MaxNights      -1.491e-03  7.393e-04  -2.016 0.043784 *
## ---
## Signif. codes:
## 0 '***' 0.001 '**' 0.01 '*' 0.05 '.' 0.1 ' ' 1
##
```

```
## Residual standard error: 40.32 on 10357 degrees of freedom
## Multiple R-squared:  0.4764, Adjusted R-squared:  0.4758
## F-statistic: 785.3 on 12 and 10357 DF,  p-value: < 2.2e-16
```

(d) All three methods come up with the same solution. While it may seem that the algorithms should come to the same solution, it is not always the case.

9.12   Best Subsets Regression

The best subsets regression algorithm consists of evaluating all possible subsets. Therefore, best subsets regression guarantees that the chosen model will be the *best* model according to the criteria that you choose. For instance, the best subsets will find the model with the best adjusted R^2 for all variable combinations if that is our criteria.

The implementation of best subsets regression in R is given within the leaps package, and the function that invokes the algorithm is regsubsets. The regsubsets requires a formula and the data set. An optional argument of nvmax specifies the number of variables to be considered. By default, regsubsets considers variable combinations up to 8. Best subsets regression can be coded using the formula fmla and data df:

```
library(leaps)
BSR = regsubsets(fmla, data = df)
```

9.13   Airbnb Pricing Application: Best Subsets Regression 1

Using the Edinburgh Airbnb data file, do the following:

(a) Run the best subsets method on the data using all predictor variables using the default value of nvmax.
(b) Plot the results of the best subsets method from part (a) using BIC as the scale. Repeat using adjusted R^2 as the scale.
(c) Interpret the plots from part (b) by specifying the variables used.
(d) Find the number of variables that correspond to the best BIC, adjusted R^2, Mallows's C_p, and SSE.

Solution

(a) Here we load in the leaps package using the library function. The best subsets object, BSR, is created using the regsubsets function. The primary input arguments for this function are the same as those of the lm function. By only setting the primary input arguments, the default argument of nvmax of 8 is used.

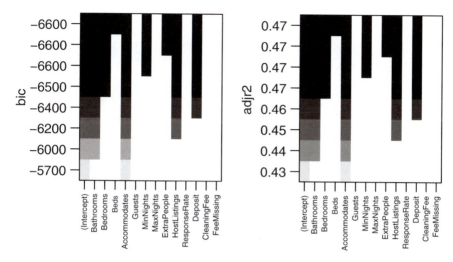

Fig. 9.4 Best subsets regression results 1

```
library(leaps)
BSR = regsubsets(Price ~ ., data = df)
```

(b) We can simply pass BSR to the `plot` function to get a best subsets plot with
    BIC as the scale. To use adjusted R^2 as the scale, set the `scale` argument to
    "adjr2" in the `plot` function.

```
par(mfrow=c(1,2))
plot(BSR)
plot(BSR, scale = "adjr2")
```

(c) In both cases, the variables returned are Bathrooms, Bedrooms, Beds,
    Accommodates, MinNights, ExtraPeople, HostListings, and Deposit.
    This information can be seen from the top row of the plots in Fig. 9.4. It is also
    possible to list out these variables using the `coef` function. The arguments of
    this function are BSR and the number of predictor variables to be used.

```
coef(BSR,8)
```

```
##   (Intercept)      Bathrooms        Bedrooms            Beds
##    3.16272079    12.73168880      8.98549616      3.26311019
## Accommodates       MinNights     ExtraPeople HostListings
##   12.87694620     2.24713688     -0.18647225      0.38641557
##       Deposit
##    0.06450636
```

(d) For convenience, we create the BSR_summary object as the summary of the
    BSR object. The BSR_summary$adjr2 consists of the best adjusted R^2 for each
    number of variables used, and each index in BSR_summary$adjr2 corresponds
    to the number of variables used.

```
BSR_summary = summary(BSR)
BSR_summary$adjr2
```

```
## [1] 0.4250282 0.4394487 0.4513901 0.4620832 0.4694081
## [6] 0.4716762 0.4731864 0.4743464
```

By inspection of the output, we see that the last value is the highest adjusted R^2 value. Since
the number of variables corresponds to the index of the vector in BSR_summary$adjr2, the
best adjusted R^2 corresponds to using 8 variables. To find the number of variables of the
highest adjusted R^2 in a programmatic way, we use the which.max function since it returns
the index number of the maximum value. Since we would like the lowest BIC, Mallows's
C_p, and SSE, we use which.min to return the best number of variables for each metric.

```
which.max(BSR_summary$adjr2)
which.min(BSR_summary$bic)
which.min(BSR_summary$cp)
which.min(BSR_summary$rss)
```

```
## [1] 8
```

```
## [1] 8
```

```
## [1] 8
```

```
## [1] 8
```

9.14   Airbnb Pricing Application: Best Subsets Regression 2

Using the Edinburgh Airbnb data file, do the following:

(a) Run the best subsets method on the data using all predictor variables using the
    nvmax option to include all possible combinations of predictor variables.
(b) Plot the results from the previous part.
(c) Find the number of variables that result in the best adjusted R^2, BIC, Mallows's
    C_p, and SSE. Also, find the best adjusted R^2, BIC, Mallows's C_p, and SSE.
(d) Plot the adjusted R^2 as a function of the number of variables. Repeat for BIC,
    Mallows's C_p, and SSE.
(e) What are the advantages and disadvantages of each model selected from the
    previous part?

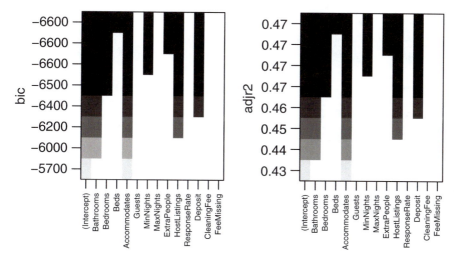

Fig. 9.5 Best subsets regression results 2

Solution

(a) Here, we again use `regsubsets` in a similar fashion to part (a) in the previous application but specify `nvmax` to include all 13 variables.

```
BSR2 = regsubsets(Price ~ ., data = df, nvmax = 13)
```

(b) In Fig. 9.5, we plot a side-by-side plot of the best subsets plots using a BIC scale and an adjusted R^2 scale.

```
par(mfrow=c(1,2))
plot(BSR)
plot(BSR, scale = "adjr2")
```

(c) For convenience, we define the BSR_summary2 summary object.

```
BSR_summary2 = summary(BSR2)
```

The values of the adjusted R^2, BIC, Mallows's C_p, and SSE are printed using the `max` and `min` functions here.

```
adjr2 = max(BSR_summary2$adjr2)
bic = min(BSR_summary2$bic)
cp = min(BSR_summary2$cp)
sse = min(BSR_summary2$rss)

adjr2
```

```
bic
cp
sse
```

```
## [1] 0.475788
```

```
## [1] -6598.668
```

```
## [1] 12.9348
```

```
## [1] 16837477
```

A number of variables corresponding to the best adjusted R^2, BIC, Mallows's C_p, and SSE are found using the which.max and which.min functions here.

```
nvar_adjr2 = which.max(BSR_summary2$adjr2)
nvar_bic = which.min(BSR_summary2$bic)
nvar_cp = which.min(BSR_summary2$cp)
nvar_sse = which.min(BSR_summary2$rss)

nvar_adjr2
nvar_bic
nvar_cp
nvar_sse
```

```
## [1] 12
```

```
## [1] 9
```

```
## [1] 12
```

```
## [1] 13
```

Plotting the metrics in a 2 by 2 grid can be accomplished using the par command. The plot in the upper left corner of Fig. 9.6 shows the vector BSR_summary2$adjr2 plot with the number of variables on the x-axis. The plot command was used here with a vector which results in the vector indices to be the values on the x-axis. The xlab and ylab arguments are used to label the axes and we further specify the type option to be a line rather than points. After using the plot function, we use the points function which places points on the previous plot. In the case of the adjusted R^2 plot, a red point is inserted at $(12, 0.4758)$ which denotes the highest adjusted R^2. The optional argument col is specified to make this point red. The process is repeated for BIC, Mallows's C_p, and SSE.

```
par(mfrow=c(2,2))

# Adjusted R-Squared
plot(BSR_summary2$adjr2, xlab = "Number of Variables",
     ylab = "Adj. R-Squared", type = "l")
```

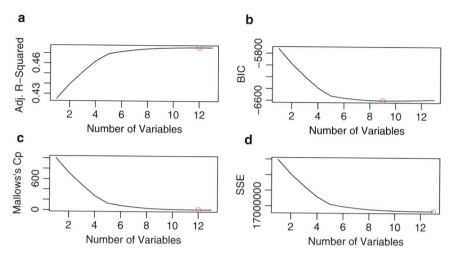

Fig. 9.6 Best subsets regression metric plots

```
points(nvar_adjr2, adjr2, col = "red")

# BIC
plot(BSR_summary2$bic, xlab = "Number of Variables",
     ylab = "BIC", type = "l")
points(nvar_bic, bic, col = "red")

# Mallows's Cp
plot(BSR_summary2$cp, xlab = "Number of Variables",
     ylab = "Mallows's Cp", type = "l")
points(nvar_cp, cp, col = "red")

# SSE
plot(BSR_summary2$rss, xlab = "Number of Variables",
     ylab = "SSE", type = "l")
points(nvar_sse, sse, col = "red")
```

(e) In the models fit using best subsets, the model with the best fit is the model that includes all the variables since it has the lowest SSE. However, the model with the lowest error is not necessarily the best choice since it could include predictor variables that are not meaningful for the prediction. This was similar to the case with the multiple R^2, where the multiple R^2 always increased with the addition of another predictor regardless of if the predictor was actually related to the response. We were then able to modify the multiple R^2 by taking into account the number of predictors using the adjusted R^2 to get a better measure of fit.

Both adjusted R^2 and Mallows's C_p agree that the best subset of predictor variables is that with 12 predictor variables. Recall for the first application from the chapter that the variable with low significance was removed from the model. While the adjusted R^2 did not noticeably change due to the number of digits displayed in the summary, the best subsets results indicate that the model did in fact improve the adjusted R^2. Furthermore, having one less predictor variable especially one with low significance results in a more parsimonious model.

The BIC, on the other hand, was lowest when there were 9 predictor variables. The BIC measure specifies a smaller model since BIC becomes higher with more predictor variables. Arguably, this is an attractive feature since the model becomes more easily interpretable with fewer predictors.

9.15 Stepwise and Best Subsets Regression

The backward elimination, forward selection, and stepwise algorithms selected variables in an iterative manner but did not consider all of the possible predictor variable subsets. While the aforementioned iterative procedures are sufficient for most cases, in cases of many variables, the results may be insufficient since there is no guarantee that the resulting model from these methods will be the best method. Best subsets regression, on the other hand, will guarantee that the best possible metric is found across all predictor variable combinations. We note that the best subsets regression can be limited in R using the nvmax argument. This argument can be adjusted as discussed in the previous application, and however, the number of subset models needed to be fit grows at a rapid rate as the number of variables gets large. Since the number of variable subsets becomes very large when the number of variables is large, the best subsets regression algorithm can be computationally expensive. In the Airbnb applications, all of the iterative approaches had the same resulting model which would also yield the same model for best subsets regression using the AIC metric (this exercise is left to the reader).

Note that the selection methods from this chapter are only a starting place for what variables should be included. After determining a good subset of variables, the analyst should continue their analysis to see if interaction or higher order terms should be included in the final model.

9.16 Case Study: Cancer Treatment Cost Analysis

9.16.1 Problem Statement

An insurance company heavily relies on various factors to determine what an individual's premium should be. A profitable insurance company charges more in premiums, on average, than the amount of health care expenses incurred by beneficiaries. Hence, insurance company analysts must study the profiles of individual beneficiaries to properly estimate the future expenses that each may incur and then properly calculate the appropriate premium for that individual.

In this case study, we focus on the factors of health care costs that can be used by an insurance company to determine premiums. Health care costs can vary for a variety of reasons. Among these, cancer treatment represents a particularly costly medical expense that insurance pays for. This case examines the data on charges billed by health insurance for the treatment of different types of cancer.

Carla is a junior analyst at Quality Insurance (QI). She was recently tasked with examining insurance charges made by policyholders who were diagnosed with cancer. She fit a regression model from the data, but some issues regarding the model need to be resolved. After Carla finished her analysis, you were hired as an independent contractor to check the quality of her analysis and answer a few additional questions. Specifically, your tasks consist of the following:

1. Assess the regression model fit from the preliminary analysis.
2. Find a parsimonious model that predicts insurance charges and interpret the coefficients.
3. Test for differences in charges across the different BMI categories for smokers and nonsmokers.
4. Test for income differences across the different insurance plans.

9.16.2   Data Description

QI provides data concerning cancer patients from 2021. Several variables included within the data can be used to predict the amount of charges billed to QI.

The data set consists of the variables listed below:

- ID—The row or observation number, where each row represents an individual.
- Age—The age of the individual receiving treatment.
- BMI—Body Mass Index (BMI), a measure of an individual's weight in reference to their height.
- Income—The income of the individual.
- Plan—A categorical variable representing one of the three types of medical insurance plans offered by QI.
- Smoke—A categorical variable representing whether an individual smokes or not.
- Sex—A categorical variable representing whether an individual is male or female.
- Charges—Medical costs billed to health insurance.

The data is contained within the Charges.csv file.

9.16.3   Preliminary Analysis

Carla performed a preliminary analysis. She loaded in the data and fit a regression model using all of the columns, other than `Charges`, as predictors. Her code and summary are shown here.

```
df = read.csv("Charges.csv")
reg = lm(Charges ~ ., data = df)
summary(reg)
```

```
##
## Call:
## lm(formula = Charges ~ ., data = df)
##
## Residuals:
##     Min      1Q  Median      3Q     Max
## -99886  -26009   -3379   16961  222706
##
## Coefficients:
##                  Estimate Std. Error t value Pr(>|t|)
## (Intercept) -7.486e+04  1.753e+04  -4.269 2.15e-05 ***
## ID          -6.967e+00  4.806e+00  -1.450   0.1475
## Age          9.403e+02  1.893e+02   4.968 7.96e-07 ***
## BMI          3.979e+03  3.072e+02  12.953  < 2e-16 ***
## Plan         2.190e+03  2.316e+03   0.946   0.3445
## Sexmale      4.913e+03  2.797e+03   1.757   0.0793 .
## SmokeY       5.284e+04  3.386e+03  15.607  < 2e-16 ***
## Income       1.290e-01  7.222e-02   1.786   0.0745 .
## ---
## Signif. codes:
## 0 '***' 0.001 '**' 0.01 '*' 0.05 '.' 0.1 ' ' 1
##
## Residual standard error: 43820 on 992 degrees of freedom
## Multiple R-squared:  0.3006, Adjusted R-squared:  0.2957
## F-statistic:  60.9 on 7 and 992 DF,  p-value: < 2.2e-16
```

Unfortunately, there are several errors in the analysis. The residual plots indicate that the linear regression assumptions have been violated.

```
par(mfrow = c(1, 2))
hist(reg$residuals, xlab = "Residuals",
     main = "Histogram of Residuals")
plot(reg$fitted.values, reg$residuals, xlab = "Fitted values",
     ylab = "Residuals", main = "Residual Plot")
```

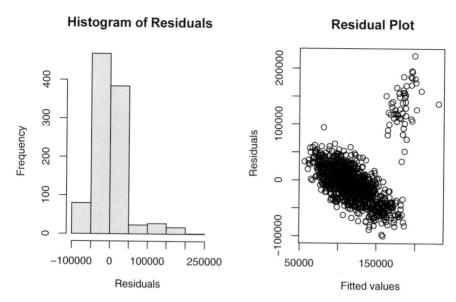

Fig. 9.7 Diagnostic plots for preliminary model

On the left-side of Fig. 9.7, the residuals clearly show a pattern that changes as the fitted values increase. The pattern of the residuals violates the assumption of linearity. On the right-side of Fig. 9.7, the bell-shaped pattern of a histogram is not evident since the plot has a right skew indicating the regression model is non-normal, which is another violation of the linear regression assumptions. Here we use the shapiro.test function to invoke the Shapiro-Wilk test.

```
shapiro.test(reg$residuals)
```

```
##
##  Shapiro-Wilk normality test
##
## data:  reg$residuals
## W = 0.88448, p-value < 2.2e-16
```

The results of the Shapiro-Wilk test further verify the claim that the residuals follow a non-normal distribution. In particular, the Shapiro-Wilk p-value is well below 5% indicating that the distribution of the residuals proves to be non-normal. However, the coefficients are still the best estimates for the linear model, but further exploration is warranted.

9.16.4   Revised Analysis

After the preliminary analysis, some of Carla's senior employees advise you that more careful inspection of the variables is required before fitting a regression model. Several functions in R exist to observe the data. Here we will make use of the summary function.

```
summary(df)
```

```
##          ID                 Age                 BMI
##   Min.   :    1.0   Min.   :49.00   Min.   :16.13
##   1st Qu.: 250.8   1st Qu.:58.00   1st Qu.:23.77
##   Median : 500.5   Median :66.00   Median :26.80
##   Mean   : 500.5   Mean   :66.03   Mean   :26.90
##   3rd Qu.: 750.2   3rd Qu.:74.00   3rd Qu.:30.14
##   Max.   :1000.0   Max.   :82.00   Max.   :40.33
##          Plan               Sex                 Smoke
##   Min.   :1.000   Length:1000       Length:1000
##   1st Qu.:1.750   Class :character   Class :character
##   Median :2.000   Mode  :character   Mode  :character
##   Mean   :2.074
##   3rd Qu.:3.000
##   Max.   :3.000
##         Income             Charges
##   Min.   : 34157   Min.   : 44071
##   1st Qu.: 62055   1st Qu.: 96108
##   Median : 87953   Median :110874
##   Mean   : 86569   Mean   :120040
##   3rd Qu.:107296   3rd Qu.:126588
##   Max.   :165321   Max.   :421281
```

Notice that Smoke and Sex are not listed as factor variables. While these variables can be manually converted to factors as shown in Chap. 5, it is simpler to make the conversion when loading in the data. By setting the stringsAsFactors option to TRUE, within the read.csv command, the text string vectors are automatically converted to factor variables. Within this data set, converting all of the string vectors to factor variables is an elegant solution for this data, since all of the non-numeric variables are categorical. If a date column existed within the data, the stringsAsFactors option should not be set to TRUE, since a date does not typically denote a category level.

```
df = read.csv("Charges.csv", stringsAsFactors = TRUE)
```

Furthermore, from the summary and the description of the variables, it is evident that the ID column should not be used as a predictor variable. Setting the value of ID to NULL will remove the column from the data frame.

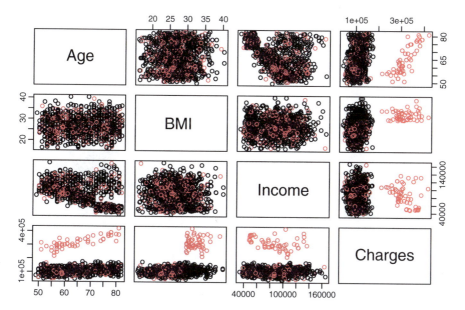

Fig. 9.8 Scatterplot matrix colored by smokers

```
df$ID = NULL
```

We noted in the variable explanation that the Plan variable is categorical. Since the Plan variable takes on three numeric values (1, 2, and 3), R incorrectly assumes that Plan is numeric. Using the factor function converts Plan to a factor variable.

```
df$Plan = factor(df$Plan)
```

An initial inspection of the variables can be done by finding a scatterplot matrix among the numeric variables. Using the subset function, we can create a data frame of numeric values named numeric_var, as shown in the code below:

```
numeric_var = subset(df, select = c("Age", "BMI", "Income",
                                    "Charges"))
```

Interesting scatterplot matrices can be created by coloring the variables by categorical variables. In this instance, the Smoke variable is used to color the data by smokers and nonsmokers. Here we use the plot function and specify the col argument as the Smoke variable within df.

```
plot(numeric_var, col = df$Smoke)
```

In Fig. 9.8, red represents smokers and black represents nonsmokers. Because smoking is a serious health risk, the charges incurred are much greater for smokers as shown.

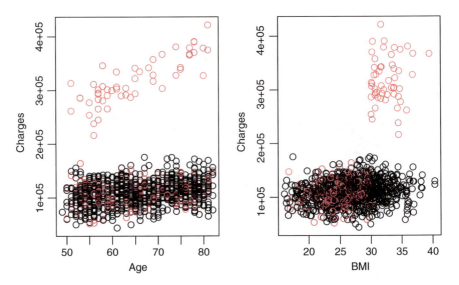

Fig. 9.9 Scatterplots

Note that a similar set of scatterplot matrices can be created by using the numeric variables and coloring them by Sex and Plan. The code to create the plots is similar to that of the previous plot command.

```
plot(numeric_var, col = df$Sex)
plot(numeric_var, col = df$Plan)
```

The scatterplot matrices are not displayed here because they lack significance. The analyst should create and inspect the scatterplots to determine their relevance.

After observing the scatterplots colored by smoker, insurance charges notably differ depending on the customer's status as a smoker or a nonsmoker, as evidenced by the relationship between Age and Charges. A larger scatterplot between Age and Charges on the right-side of Fig. 9.9 reveals more details about this relationship. The scatterplots also reveal a relationship between BMI and Charges, which also warrants further investigation and, therefore, is shown on the left side of Fig. 9.9.

```
par(mfrow = c(1, 2))
plot(Charges ~ Age, data = df, col = df$Smoke)
plot(Charges ~ BMI, data = df, col = df$Smoke)
```

Investigating the left-side of Fig. 9.9, we see that there is a linear relationship between Age and Charges for nonsmokers as indicated by the black points. Above the black points, there is a linear trend consisting of red points. Thus, it becomes clear that Charges and Age may have a different relationship for smokers and nonsmokers.

Table 9.1 Weight status by body mass index

Body mass index (BMI)	Weight status
BMI < 18.5	Underweight
$18.5 \leq$ BMI < 25	Normal
$25 \leq$ BMI < 30	Overweight
BMI $\geq$ 30	Obese

Investigating the right-side of Fig. 9.9, we find the divide between the `Charges` that smokers have when their `BMI` is lower or higher than 30. On the BMI scale, a value of 30 represents the threshold between overweight and obese, which is a health risk that could cause medical charges to be higher. The weight status categories from BMI levels are given in Table 9.1. From Fig. 9.9, we see clearly that a dummy variable for obese individuals could be included to more accurately model `Charges`.

The following line of code creates a dummy variable based on the numeric `BMI` column. The result of the inequality test is either `"Y"` or `"N"` for each observation.

```
df$Obese = ifelse(df$BMI > 30, "Y", "N")
```

While the BMI scale is relatively well known, it may come as a surprise to some that the previous scatterplots show different trends when BMI is above 30. Our analysis therefore prompts the insurance company to identify obese individuals, who are more at risk and will most likely have more insurance charges.

9.16.5 Regression Modeling

From the analysis in the previous section, we note that smokers should be treated differently in modeling their medical charges to account for the effects of smoking. We achieve this goal by using interactions. Here, we would like to check the models consisting of all predictor variables, as well as the variable interactions with smokers. Thus, the interaction terms are used:

- `Age:Smoke`
- `BMI:Smoke`
- `Income:Smoke`
- `Obese:Smoke`.

In addition, we would like to consider the effects of obese smokers since they are at a higher risk of incurring charges. The three-way interactions are included:

- `Age:Obese:Smoke`
- `BMI:Obese:Smoke`
- `Income:Obese:Smoke`

For convenience, the formula consisting of `Charges` as predicted by all other variables and the interactions is defined and named `fmla`.

```
fmla = Charges ~ . + Age:Smoke + BMI:Smoke + Income:Smoke +
  Obese:Smoke + Age:Obese:Smoke + BMI:Obese:Smoke +
  Income:Obese:Smoke
```

Using the formula given by `fmla`, a full model is fit and summarized.

```
full = lm(fmla, data = df)
summary(full)
```

```
##
## Call:
## lm(formula = fmla, data = df)
##
## Residuals:
##    Min     1Q Median    3Q    Max
## -69239 -13645    442  13449  64956
##
## Coefficients:
##                         Estimate Std. Error t value Pr(>|t|)
## (Intercept)            4.243e+04  1.215e+04   3.493 0.000498 ***
## Age                    6.346e+02  1.166e+02   5.443 6.61e-08 ***
## BMI                    8.975e+02  2.624e+02   3.421 0.000651 ***
## Plan2                 -1.620e+03  1.646e+03  -0.985 0.325069
## Plan3                  1.090e+04  2.252e+03   4.840 1.51e-06 ***
## Sexmale                3.169e+03  1.319e+03   2.403 0.016450 *
## SmokeY                -4.252e+03  2.384e+04  -0.178 0.858497
## Income                -3.901e-02  4.584e-02  -0.851 0.394996
## ObeseY                 2.361e+04  2.835e+04   0.833 0.405159
## Age:SmokeY            -5.844e+01  2.351e+02  -0.249 0.803753
## BMI:SmokeY             1.468e+02  5.892e+02   0.249 0.803303
## SmokeY:Income          6.186e-02  7.732e-02   0.800 0.423871
## SmokeY:ObeseY         -2.572e+04  6.450e+04  -0.399 0.690130
## Age:SmokeN:ObeseY     -8.561e+01  2.101e+02  -0.407 0.683772
## Age:SmokeY:ObeseY      3.783e+03  4.499e+02   8.409  < 2e-16 ***
## BMI:SmokeN:ObeseY     -6.943e+02  7.073e+02  -0.982 0.326507
## BMI:SmokeY:ObeseY     -1.909e+03  1.462e+03  -1.306 0.192010
## SmokeN:Income:ObeseY   3.897e-02  6.647e-02   0.586 0.557820
## SmokeY:Income:ObeseY   2.398e-01  1.539e-01   1.558 0.119449
## ---
## Signif. codes:  0 '***' 0.001 '**' 0.01 '*' 0.05 '.' 0.1 ' ' 1
##
## Residual standard error: 20520 on 981 degrees of freedom
## Multiple R-squared:  0.8483, Adjusted R-squared:  0.8455
## F-statistic: 304.7 on 18 and 981 DF,  p-value: < 2.2e-16
```

The `full` model above returns a great fit, especially compared with the low
adjusted R^2 from the initial model. However, the model is not easily interpreted
since there are too many predictor variables. In the following section, other models
will be considered.

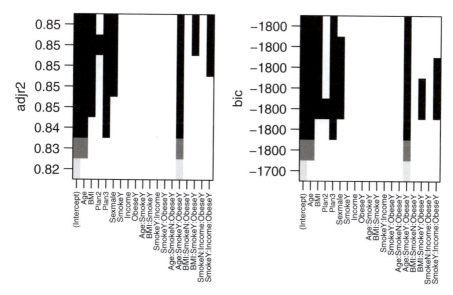

Fig. 9.10 Best subsets plots

Best Subsets Model

Note that the model above has a significantly higher adjusted R^2 than the model fit in the preliminary analysis. However, the model includes many predictors which should not be used. To find a more parsimonious model, we can utilize best subsets. In particular, the leaps package allows us to use the regsubsets function.

```
library(leaps)
bsr = regsubsets(fmla, data = df)
```

The results from the best subsets object, bsr, can be observed by using the plot function. Here, we plot the models ranked according to their adjusted R^2 value on the left-side of Fig. 9.10 and ranked according to their best BIC values on the right-side.

The coefficients of the best subset model can be found using the coef function. However, this function does not return the complete summary. While the coef function allows you to find the best model for different numbers of variables, here we choose four variables since the plot in the right-side of Fig. 9.10 has the lowest BIC.

```
coef(bsr, 4)
```

```
##      (Intercept)             Age            BMI
##       44878.9832        614.8044       740.3576
##            Plan3 Age:SmokeY:ObeseY
##       11650.6970        3163.9707
```

To get a summary of the model chosen from best subsets, the lm function can be used with the summary function.

```
reg = lm(Charges ~ Age + BMI + Plan + Age:Smoke:Obese,
         data = df)
summary(reg)
```

```
##
## Call:
## lm(formula = Charges ~ Age + BMI + Plan + Age:Smoke:Obese,
##     data = df)
##
## Residuals:
##    Min     1Q Median     3Q    Max
## -70833 -13994    102  13566  65223
##
## Coefficients: (1 not defined because of singularities)
##                       Estimate Std. Error t value Pr(>|t|)
## (Intercept)           43880.69    7556.47   5.807 8.56e-09 ***
## Age                    3774.01      87.45  43.156  < 2e-16 ***
## BMI                     809.95     216.44   3.742 0.000193 ***
## Plan2                 -1521.51    1639.40  -0.928 0.353585
## Plan3                 10671.44    1733.82   6.155 1.09e-09 ***
## Age:SmokeN:ObeseN     -3156.16      51.67 -61.078  < 2e-16 ***
## Age:SmokeY:ObeseN     -3146.27      56.20 -55.986  < 2e-16 ***
## Age:SmokeN:ObeseY     -3169.67      48.17 -65.802  < 2e-16 ***
## Age:SmokeY:ObeseY          NA         NA      NA       NA
## ---
## Signif. codes:  0 '***' 0.001 '**' 0.01 '*' 0.05 '.' 0.1 ' ' 1
##
## Residual standard error: 20570 on 992 degrees of freedom
## Multiple R-squared:  0.8459, Adjusted R-squared:  0.8448
## F-statistic:  778 on 7 and 992 DF,  p-value: < 2.2e-16
```

While the coefficient values are relatively similar to those returned using the coef function, they are not the same since the lm function includes some variables not used in the best subsets model. For instance, the summary output includes the Plan2 and Plan3 dummy variables, whereas the best subsets results from Fig. 9.10 did not contain the Plan2 dummy variable. A character variable for Plan3 can be created using the ifelse function. This character variable will automatically be converted to a factor variable within the lm function.

```
df$Plan3 = ifelse(df$Plan == 3, "Y", "N")
```

Another difference between the results of best subsets from Fig. 9.10 and the summary results occurs as more interactions between Age and the levels of Smoke and Obese. To remedy this, we again use the ifelse statement. The Obese_Smoke dummy indicates that an observation is both obese and a smoker.

```
df$Obese_Smoke = ifelse(df$Obese == "Y" & df$Smoke == "Y", "Y",
                  "N")
```

Using the manually created variables within the lm function allows us to print the summary below:

```
reg = lm(Charges ~ Age + BMI + Plan3 + Age:Obese_Smoke,
         data = df)
summary(reg)
```

```
##
## Call:
## lm(formula = Charges ~ Age + BMI + Plan3 + Age:Obese_Smoke,
##     data = df)
##
## Residuals:
##     Min     1Q Median     3Q     Max
## -70677 -14163    148  13410  64667
##
## Coefficients:
##                     Estimate Std. Error t value Pr(>|t|)
## (Intercept)         44878.98    6231.85   7.202 1.17e-12 ***
## Age                   614.80      72.47   8.483  < 2e-16 ***
## BMI                   740.36     150.90   4.906 1.08e-06 ***
## Plan3Y              11650.70    1389.61   8.384  < 2e-16 ***
## Age:Obese_SmokeY     3163.97      46.45  68.118  < 2e-16 ***
## ---
## Signif. codes:  0 '***' 0.001 '**' 0.01 '*' 0.05 '.' 0.1 ' ' 1
##
## Residual standard error: 20550 on 995 degrees of freedom
## Multiple R-squared:  0.8457, Adjusted R-squared:  0.8451
## F-statistic:  1364 on 4 and 995 DF,  p-value: < 2.2e-16
```

In the summary above, the best subsets model was recreated using the lm function, which can be confirmed since the coefficients match the best subsets coefficients displayed using the coef function. Also of note, since the best model was chosen according to the best BIC value, a more parsimonious model was returned despite the lower adjusted R^2 in the best subsets model. While the summary indicates a parsimonious model, it is important to also check the diagnostic plots.

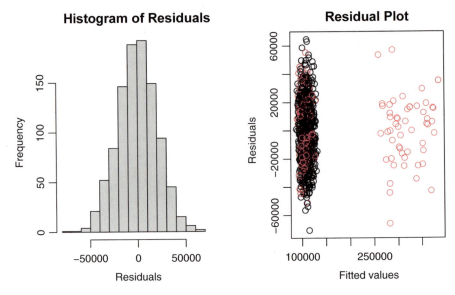

Fig. 9.11 Diagnostic plots for best subsets model

```
par(mfrow = c(1, 2))
hist(reg$residuals, xlab = "Residuals",
     main = "Histogram of Residuals")
plot(reg$fitted.values, reg$residuals, xlab = "Fitted values",
     ylab = "Residuals", main = "Residual Plot", col = df$Smoke)
```

On the left-side of Fig. 9.11, the residuals have the same spread from left to right even though the obese–smokers have larger predicted values. The right-side of Fig. 9.11 indicates that the residuals are distributed normally. By observing the residual plots of Fig. 9.11, we can conclude that the best subsets model does not violate the linear regression assumptions.

Coefficient Interpretation

Here we interpret the coefficients of the best subsets model.

- Intercept—The intercept represents the Charges when all the predictor variables are 0. While the p-value is significant, having Age and BMI set to 0 does not make sense. If Age and BMI are mean-centered, then the intercept has more intuitive meaning.

- Age—Since the *p*-value indicates significance, the interpretation is:

  The `Charges` *will increase by $614.80 for every year increase in* `Age`*, holding all other variables constant.*

- BMI—Since the *p*-value indicates significance, the interpretation is:

  The `Charges` *will increase by $740.36 for every unit increase in BMI, holding all other variables constant.*

- Plan3Y—The *p*-value indicates this variable is significant. The interpretation is:

  The `Charges` *for plan 3 will be $11,650.70 more than the other plans, holding all other variables constant.*

- Age:Obese_SmokeY—The *p*-value indicates this variable is significant. The interpretation is:

  The `Charges` *will increase by $3163.97 more for every year increase in* `Age` *for obese smokers, holding all other variables constant.*

Charges Across BMI Categories

We would now like to test for differences among `Charges` across the various BMI categories for smokers and nonsmokers. Using the `cut` command, we can segment the BMI variable according to the cutoffs as specified in the `breaks` option. In this case, an individual with a BMI of below 18.5 is placed in the "Under" category, while 18.5 to below 25 is categorized as "Normal," while 25 to under 30 is "Over," and 30 or above is "Obese." These categories are specified in the `labels` option.

```
df$Categories = cut(df$BMI, breaks = c(-Inf, 18.5, 25, 30, Inf),
                    labels = c("Under", "Normal", "Over",
                               "Obese"))
```

Using these `Categories`, we use the `boxplot` function to get a visual comparison of the smokers by `Categories`. The smokers are specified in the data frame by indexing the values where `Smoke` is "Y."

```
boxplot(Charges ~ Categories, data = df[df$Smoke=="Y",],
        col = "lightgray")
```

Notice from the boxplot in Fig. 9.12 that the `Obese` category significantly differs from the remaining categories. Therefore, there is a notable difference in `Charges` among categories. The result can be formalized with a statistical test. Here μ_1, μ_2, μ_3, and μ_4 represent the mean `Charges` of the respective `Categories` for smokers.

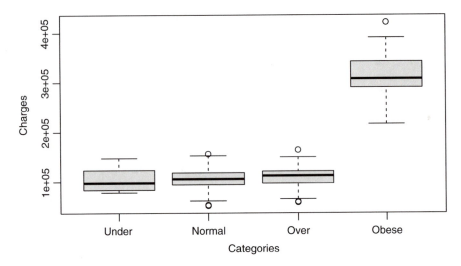

Fig. 9.12 Boxplot of charges by BMI category (smokers)

Null Hypothesis: Charges is the same across different BMI categories (for smokers).

$$H_0 : \mu_1 = \mu_2 = \mu_3 = \mu_4$$

Alternative Hypothesis: At least one mean is different.

The alternative hypothesis is equivalent to saying that there is a difference in Charges among the categories. Using the `aov` command, we calculate an ANOVA test.

```
aov1 = lm(Charges ~ Categories, data = df[df$Smoke=="Y",])
anova(aov1)
```

```
## Analysis of Variance Table
##
## Response: Charges
##             Df    Sum Sq    Mean Sq F value     Pr(>F)
## Categories   3 1.7414e+12 5.8045e+11  717.94 < 2.2e-16 ***
## Residuals  210 1.6978e+11 8.0850e+08
## ---
## Signif. codes:  0 '***' 0.001 '**' 0.01 '*' 0.05 '.' 0.1 ' ' 1
```

The *p*-value of the *F*-test is extremely small in this case indicating statistical significance. Therefore, the ANOVA test is in agreement with our assessment made by looking at the boxplots.

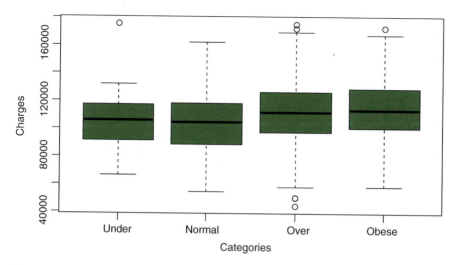

Fig. 9.13 Boxplot of charges by BMI category (nonsmokers)

Next, we analyze the relationship of nonsmokers in a similar way. The boxplot of nonsmokers by category can be colored "darkgreen" by specifying the `col` argument within the boxplot command. The nonsmokers are indexed in the data frame by selecting rows where the `Smoke` variable is set to "N."

```
boxplot(Charges ~ Categories, data = df[df$Smoke == "N",],
        col = "darkgreen")
```

From Fig. 9.13, we notice that the median does differ for each of the `Categories`, but not as drastically as they do in the smoker boxplot shown in Fig. 9.12. While the ANOVA test looks for differences in the means, the mean and the median are both measures of the center of the distribution. Here we formalize the result with a statistical test. For this analysis, we let μ_1, μ_2, μ_3, and μ_4 represent the mean `Charges` of the respective `Categories` for nonsmokers.

Null Hypothesis: Charges are the same across different BMI categories (for nonsmokers).

$$H_0 : \mu_1 = \mu_2 = \mu_3 = \mu_4$$

Alternative Hypothesis: At least one mean is different.
Using the `aov` command, an ANOVA test is calculated for nonsmokers.

```
aov2 = lm(Charges ~ Categories, data = df[df$Smoke == "N",])
anova(aov2)
```

```
## Analysis of Variance Table
##
## Response: Charges
##               Df      Sum Sq     Mean Sq F value     Pr(>F)
## Categories     3 1.1427e+10 3808930637  7.9686 3.082e-05 ***
## Residuals    782 3.7379e+11   477991129
## ---
## Signif. codes:  0 '***' 0.001 '**' 0.01 '*' 0.05 '.' 0.1 ' ' 1
```

The p-value of the F-test indicates that a difference exists among the categories for nonsmokers. Notice here that the p-value is significantly larger than the p-value for the smoker F-test, which indicates that the difference among smokers is more significant than nonsmokers. This result is consistent with our conclusion from observing the boxplots.

Income Across Plans

Our last task will consist of analyzing the three plans by `Income`, since we consider this factor to differ across the three plans.

The mean values by `Plan` can be found using the by function. Note that the by function applies the mean function across the different plans. The first argument in the by function is the numeric variable from which we want to calculate the mean. The second argument is the factor variable that we want the means separated by. Lastly, we specify the function that we want applied to the different categories for the numeric variable is specified.

```
by(df$Income, df$Plan, mean)
```

While the mean income of individuals with Plans 1 and 2 is relatively similar, the mean income of individuals with Plan 3 is very different. Here we produce a boxplot to visually show the difference of `Income` among `Plans`.

```
boxplot(Income ~ Plan, data = df, col = "lightblue")
```

In Fig. 9.14, the plans correspond to various income values. Lastly, we will test for a difference using a formal F-test. We let μ_1, μ_2, and μ_3 represent the mean income corresponding to Plans 1, 2, and 3.

Null Hypothesis: Income is not statistically significant across different plans.

$$H_0 : \mu_1 = \mu_2 = \mu_3$$

Alternative Hypothesis: At least one mean is different among the plans.

As seen in Fig. 9.14, income varies by plan. Next, we have to determine if income is statistically significant across the separate plans. To do so, we need to create a regression model between `Income` and `Plan` and examine the significance of the model and individual variables.

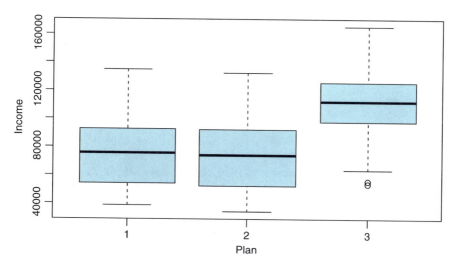

Fig. 9.14 Boxplots by plan

```
aov3 = lm(Income ~ Plan, data = df)
anova(aov3)
```

```
## Analysis of Variance Table
##
## Response: Income
##              Df     Sum Sq    Mean Sq F value    Pr(>F)
## Plan          2 3.0089e+11 1.5044e+11  328.12 < 2.2e-16 ***
## Residuals   997 4.5712e+11 4.5850e+08
## ---
## Signif. codes:  0 '***' 0.001 '**' 0.01 '*' 0.05 '.' 0.1 ' ' 1
```

This summary shows that plans are different based on the p-value of the F-statistic. In particular, this indicates that the model is statistically significant when compared to an alpha of 5%, meaning that not all plans are the same and at least one of the plans differs. For more details, a summary of the regression model is produced.

```
summary(aov3)
```

```
##
## Call:
## lm(formula = Income ~ Plan, data = df)
##
## Residuals:
##     Min      1Q Median      3Q     Max
```

```
## -57870 -19877    134  16231  59476
##
## Coefficients:
##                Estimate Std. Error t value Pr(>|t|)
## (Intercept)     75300       1354  55.603   <2e-16 ***
## Plan2           -1168       1706  -0.685    0.494
## Plan3           36315       1802  20.147   <2e-16 ***
## ---
## Signif. codes:  0 '***' 0.001 '**' 0.01 '*' 0.05 '.' 0.1 ' ' 1
##
## Residual standard error: 21410 on 997 degrees of freedom
## Multiple R-squared:  0.3969, Adjusted R-squared:  0.3957
## F-statistic: 328.1 on 2 and 997 DF,  p-value: < 2.2e-16
```

Note that the F-statistic is the same in the ANOVA output and summary. However, the summary shows that Plans 1 and 2 do not differ significantly based on the individual significance t-test of Plan2. Plan3, on the other hand, differs significantly from Plan1, which is the underlying reason for the significance of the F-statistic noted by the low p-value.

9.16.6   Recommendations and Findings

We find the following with respect to the tasks that we were given:

1. Assess the regression model fit from the preliminary analysis.

   A review of the preliminary analysis showed that the model not only violated the linear regression assumptions but also used invalid variables and neglected some important interactions.

2. Find a parsimonious model that predicts insurance charges and interpret the coefficients.

   After applying best subsets and finding the minimum BIC, a model with only three predictor variables and one interaction was able to achieve an adjusted R^2 of 0.8451. The coefficients are interpreted in the Coefficient Interpretation section above.

3. Test for differences in charges across the various BMI categories for smokers and nonsmokers.

   Both tests across the different categories for smokers and nonsmokers indicated a significant difference among categories, with the most extreme difference being that of smokers in the obese category.

4. Test for income differences across the three insurance plans.

   The test revealed that there were differences in the income for insurance plans. In particular, insurance plan 3 differs from other plans.

9.16.7 Case Conclusion

This case demonstrates how modern health care is being transformed by analytics and more specifically how analytics can provide insights from health care data using a detailed example of the process to determine health care charges. Since greater insight is known about what determines health care charges, QI's decision-makers can adjust the cost of premiums to offset these charges.

The analysis done in this case study utilized methods presented throughout this book. We used several R commands to model and analyze the data, such as anova, boxplot, by, cut, factor, hist, ifelse, lm, par, plot, read.csv, shapiro.test, subset, and summary. In addition to the base R commands, we used the regsubsets function which is within the leaps package.

Problems

1. **Health Care Costs Iterative Methods**
   Load in the data:

   ```
   df = read.csv("Charges.csv")
   ```

   Using the insurance charges data set, build a model to predict the charges considering all other variables from the data set.

   a. Perform backward elimination and produce a summary.
   b. Perform forward selection and produce a summary.
   c. Perform stepwise regression and produce a summary.
   d. Are the models the same? If not, which one is different?
   e. How well does the backward elimination model fit the data?
   f. Construct a residual plot and a histogram of the residuals of the backward elimination model and discuss. By observing the plots, is it evident if any of the linear regression model assumptions are violated?

2. **Country GDP Iterative Methods**
   Load in the data:

   ```
   df = read.csv("Countries.csv")
   ```

   Using the countries data set, build a model to predict the GDP considering all other relevant variables from the data set.

   a. Perform backward elimination and produce a summary.
   b. Perform forward selection and produce a summary.
   c. Perform stepwise regression and produce a summary.
   d. Are the models the same? If not, which one is different?

e. How well does the backward elimination model fit the data?
f. Construct a residual plot and a histogram of the residuals of the backward elimination model and discuss. By observing the plots, is it evident if any of the linear regression model assumptions are violated?

3. **Country GDP Best Subsets Regression**
Load in the data:

```
df = read.csv("Countries.csv")
```

Using the countries data set, build a model to predict the GDP considering all other relevant variables from the data set.

a. Fit a best subsets model to predict GDP considering all other relevant variables from the data set.
b. Use the plot function on the best subsets object from part (a).
c. Which variables are in the best subset?
d. Plot the best subsets object again, but include the option: scale='adjr2'.
e. Which variables are in the best subset using adjusted R-squared?
f. Print out the best adjusted R-squared to at least 4 decimal places.
g. Construct a residual plot and a histogram of the residuals of the best subsets model indicated by BIC, and discuss if the linear regression model assumptions are met. (Hint: You may want to recreate the model fit using the lm function to get the residuals.)

4. **Country GDP Best Subsets Regression 2**
Load in the data:

```
df = read.csv("Countries.csv")
```

Using the countries data set, build a model to predict the GDP considering all other relevant variables from the data set.

a. Fit a best subsets model to predict the GDP considering all other relevant variables and the interactions of all predictor variables. Hint: After removing all of the nonrelevant predictor variables from the data, use the formula "GPD ~ .^2" to include all of the two variable interactions.
b. Use the plot function on the best subsets object from part (a).
c. Which variables are in the best subset?
d. Plot the best subsets object again, but include the option: scale='adjr2'.
e. Which variables are in the best subset using adjusted R-squared?
f. How does the model fit compare with the best model from the previous problem?

5. Airbnb Best Subsets Regression

Load in the data:

```
df = read.csv("AirBnb.csv")
```

Using the Airbnb data set, build a model to predict the `Price` considering all other relevant variables from the data set.

a. Fit a best subsets model to predict `Price` considering all other relevant variables and the interactions of all predictor variables. Hint: After removing all of the nonrelevant predictor variables from the data, use the formula `"Price ~ .^2"` to include all of the two variable interactions.
b. Use the plot function on the best subsets object from part (a).
c. Which variables are in the best subset?
d. Plot the best subsets object again, but include the option: scale='adjr2'.
e. Which variables are in the best subset using adjusted R-squared?
f. How does the model fit compare with the best model from the previous problem?

Appendix A
Installing Packages

This appendix chapter will guide the reader through installing packages in R. Generally speaking, there are two different steps necessary to use packages in R:

- Install the package.
- Load package.

We first cover the concept of packages in R. Next, we discuss how to install packages using the `install.packages` command followed by a discussion on a few other convenient methods. While we motivate this appendix chapter by installing the `ggplot2` package, this method can be used to install other packages as well.

A.1   Installation of the ggplot2 Package

As with any package outside of base R, you will need to install and download the package. The `install.packages` command does just that. Please note that you only need to use the install.packages function once to install the required packages. Thus, it's not advisable to run the `install.packages` function every time you run R code.

Here we install the `ggplot2` package using the following code:

```
install.packages('ggplot2')
```

© The Author(s), under exclusive license to Springer Nature Switzerland AG 2023
D. P. McGibney, *Applied Linear Regression for Business Analytics with R*,
International Series in Operations Research & Management Science 337,
https://doi.org/10.1007/978-3-031-21480-6

A.2   Loading in the ggplot2 Package

Once a package is installed, the library function can be used to load the package and also attach any required functions needed to run the packages. Loading in a package is necessary to run each time before the package is used.

```
library(ggplot2)
```

After running the command above, ggplot2 is ready to be used.

A.3   Additional Installation Methods

There are a few different ways of installing packages in R. If you were to load in a file that attempted to load in a package using the library command and the package was unavailable, a yellow ribbon will prompt you to install the package. This is a convenient method for installing packages when using Posit. Alternatively, in Posit, you can click on the View menu and select Show Packages. Showing the packages opens up a list of packages available and also gives the option to install new packages.

Appendix B
The quantmod Package

In this appendix chapter, we will guide the reader through the steps to download and compute the stock data for the stock beta case study. The downloaded data are available in the "Betas.csv" file.

B.1 Data Source

There is an abundance of stock data available, but it can be challenging to find a specific dataset that matches your desired criteria. For stock price data, Yahoo! Finance, http://finance.yahoo.com, provides access to thousands of stocks and indices. Additionally, accessing and downloading a csv file of stock price data from Yahoo Finance is quite simple.

B.2 Downloading a Single Stock or Index

First, simply specify a start date and an end date by creating text strings:

```
start_date = "2018-08-01"
end_date = "2021-09-01"
```

Then use the getSymbols function within the quantmod library. Note that it may be necessary to install the quantmod package the first time that an installation of R attempts to access it through the library function.

To access the S&P 500, which has symbol "SPY," one would run

```
library(quantmod)
getSymbols("SPY", from = start_date, to = end_date,
           src = "yahoo")
```

```
## [1] "SPY"
```

Notice that running the above command produces a "SPY" data set in the global environment. For simplicity and in an effort to reduce errors, the data set is already named.

Here are the first 6 observations returned using the head function:

```
head(SPY)
```

```
##            SPY.Open SPY.High SPY.Low SPY.Close SPY.Volume
## 2018-08-01  281.56   282.13  280.13   280.86   53853300
## 2018-08-02  279.39   282.58  279.16   282.39   63426400
## 2018-08-03  282.53   283.66  282.33   283.60   53935400
## 2018-08-06  283.64   284.99  283.20   284.64   39400900
## 2018-08-07  285.39   286.01  285.24   285.58   43196600
## 2018-08-08  285.39   285.91  284.94   285.46   42114600
##            SPY.Adjusted
## 2018-08-01   260.0621
## 2018-08-02   261.4789
## 2018-08-03   262.5993
## 2018-08-06   263.5623
## 2018-08-07   264.4326
## 2018-08-08   264.3216
```

The variable names in the "SPY" data set are found using the names function:

```
names(SPY)
```

```
## [1] "SPY.Open"    "SPY.High"      "SPY.Low"    "SPY.Close"
## [5] "SPY.Volume"  "SPY.Adjusted"
```

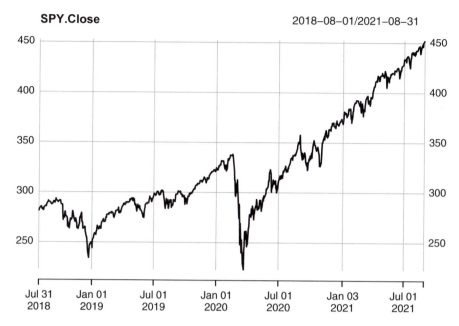

Fig. B.1 Plot of SPY

B.3   Plotting Stock Prices

To get a plot of the close prices of the S&P 500, use the plot function (Fig. B.1).

```
plot(SPY[,"SPY.Close"], main = "SPY.Close")
```

The chartSeries function within the quantmod package can also be used to plot stock data. This function has several chart variations and customizations including the use of technical indicators.

B.4   Multiple Stock Download

Download the data by inputting a vector of text strings into the getSymbols function. Also, specify the periodicity as 'monthly'.

```
stocks = c("AAPL", "CAT", "JNJ", "MCD", "PG", "MSFT", "XOM",
        "SPY")
getSymbols(stocks, from = start_date, to = end_date,
        src = "yahoo", periodicity = 'monthly')
```

```
## [1] "AAPL" "CAT"  "JNJ"  "MCD"  "PG"   "MSFT" "XOM"  "SPY"
```

B.5   Calculate Stock Returns

Using the Delt function, the adjusted stock price returns can be returned. The first
6 values are

```
head(Delt(AAPL$AAPL.Adjusted))
```

```
##               Delt.1.arithmetic
## 2018-08-01                   NA
## 2018-09-01         -0.004824849
## 2018-10-01         -0.030477482
## 2018-11-01         -0.184044545
## 2018-12-01         -0.113616477
## 2019-01-01          0.055153880
```

Note that the first return value can be calculated as follows:

```
Adjusted = data.frame(AAPL$AAPL.Adjusted)
(Adjusted[2,1] - Adjusted[1,1])/Adjusted[1,1]
```

```
## [1] -0.004824849
```

B.6   Create a Dataframe

With the data.frame and Delt functions, a dataframe of weekly returns is
generated:

```
df = data.frame(Delt(AAPL$AAPL.Adjusted),
                Delt(CAT$CAT.Adjusted),
                Delt(JNJ$JNJ.Adjusted),
                Delt(MCD$MCD.Adjusted),
                Delt(PG$PG.Adjusted),
                Delt(MSFT$MSFT.Adjusted),
                Delt(XOM$XOM.Adjusted),
                Delt(SPY$SPY.Adjusted))
names(df)=stocks
```

The first row of returns cannot be calculated with the data downloaded, so the
first observation is removed.

```
df = df[-1,]
```

Finally, the dataframe can be output to a csv using the `write.csv` command.

```
write.csv(df, "Betas.csv")
```

The "Betas.csv" file created here is used throughout the case study for this chapter.

The `BatchGetSymbols` library provides another elegant package for downloading data from multiple stock tickers. More information on this package is available at https://cran.r-project.org/.

Bibliography

D.R. Anderson, D.J. Sweeney, T.A. Williams, J.D. Camm, and J.J. Cochran. *Statistics for Business & Economics*. Cengage Learning, 2016.

Joshua D. Angrist and Jörn-Steffen Pischke. *Mostly Harmless Econometrics: An Empiricist's Companion*. Princeton University Press, December 2008.

S. Bansal. Fortune global 2000 companies (2021). https://www.kaggle.com/datasets/shivamb/fortune-global-2000-companies-till-2021, 2022.

Borle, S. Course notes for linear regression for business statistics. https://www.coursera.org/learn/linear-regression-business-statistics, 2017.

C.H. Brase and C.P. Brase. *Understanding Basic Statistics, Enhanced*. Cengage Learning, 2016.

Central Intelligence Agency. The world factbook 2021. https://www.cia.gov/the-world-factbook/, 2021.

S. Chatterjee and A.S. Hadi. *Regression Analysis by Example*. Wiley Series in Probability and Statistics. Wiley, 2006.

John Fox, Sanford Weisberg, Brad Price, Daniel Adler, Douglas Bates, Gabriel Baud-Bovy, Ben Bolker, Steve Ellison, David Firth, Michael Friendly, Gregor Gorjanc, Spencer Graves, Richard Heiberger, Pavel Krivitsky, Rafael Laboissiere, Martin Maechler, Georges Monette, Duncan Murdoch, Henric Nilsson, Derek Ogle, Brian Ripley, Tom Short, William Venables, Steve Walker, David Winsemius, Achim Zeileis, and R-Core. *R package 'car': Companion to Applied Regression*, 2022. (Version 3.1-1).

Francis Galton. Kinship and Correlation. *Statistical Science*, 4(2):81 – 86, 1989.

G. James, D. Witten, T. Hastie, and R. Tibshirani. *An Introduction to Statistical Learning: with Applications in R*. Springer Texts in Statistics. Springer New York, 2014.

Gareth James, Daniela Witten, Trevor Hastie, and Rob Tibshirani. *R package 'ISLR': Data for an Introduction to Statistical Learning with Applications in R*, 2021. (Version 1.4).

M.H. Kutner, C.J. Nachtsheim, and J. Neter. *Applied Linear Regression Models*. The McGraw-Hill/Irwin Series Operations and Decision Sciences. McGraw-Hill Higher Education, 2003.

A.M. Legendre. *Nouvelles méthodes pour la détermination des orbites des comètes*. Nineteenth Century Collections Online (NCCO): Science, Technology, and Medicine: 1780-1925. F. Didot, 1805.

Thomas Lumley and Alan Miller. *R package 'leaps': Regression Subset Selection*, 2020. (Version 3.1).

Penn State Department of Statistics. Course notes for stat 501: Regression methods. https://online.stat.psu.edu/stat501/, 2022.

© The Author(s), under exclusive license to Springer Nature Switzerland AG 2023
D. P. McGibney, *Applied Linear Regression for Business Analytics with R*,
International Series in Operations Research & Management Science 337,
https://doi.org/10.1007/978-3-031-21480-6

Jeffrey A. Ryan, Joshua M. Ulrich, Ethan B. Smith, Wouter Thielen, Paul Teetor, and Steve Bronder. *R package 'quantmod': Quantitative Financial Modelling Framework*, 2022. (Version 0.4.20).

M. Sullivan. *Statistics: Informed Decisions Using Data*. Sullivan Statistics Series. Pearson, 2010.

W. N. Venables and B. D. Ripley. *Modern Applied Statistics with S*. Springer, New York, fourth edition, 2002. ISBN 0-387-95457-0.

R.E. Walpole, R.H. Myers, S.L. Myers, and K. Ye. *Probability and Statistics for Engineers and Scientists*. Prentice Hall, 2012.

Taiyun Wei and Viliam Simko. *R package 'corrplot': Visualization of a Correlation Matrix*, 2021. (Version 0.92).

Hadley Wickham, Winston Chang, Lionel Henry, Thomas Lin Pedersen, Kohske Takahashi, Claus Wilke, Kara Woo, Hiroaki Yutani, Dewey Dunnington, and RStudio. *R package 'ggplot2': Create Elegant Data Visualisations Using the Grammar of Graphics*, 2022. (Version 3.4.0).

R.A. Wojtkiewicz. *Elementary Regression Modeling: A Discrete Approach*. Sage Publications, 2016.

C. Zhao. Edinburgh Airbnb data. https://www.kaggle.com/datasets/candicezhao28/edinburgh-airbnb-data, 2022.

Printed in the United States
by Baker & Taylor Publisher Services